Human Reliability Analysis
A Systems Engineering Approach with Nuclear Power Plant Applications

HUMAN
RELIABILITY
ANALYSIS

A Systems Engineering Approach With Nuclear Power Plant Applications

E. M. Dougherty, Jr.

and

J. R. Fragola
Advanced Technology Division
Science Applications International Corporation

A Wiley-Interscience Publication

JOHN WILEY & SONS
New York   Chichester   Brisbane   Toronto   Singapore

Copyright © 1988 by John Wiley & Sons, Inc.

All rights reserved. Published simultaneously in Canada.

Reproduction or translation of any part of this work beyond that permitted by Section 107 or 108 of the 1976 United States Copyright Act without the permission of the copyright owner is unlawful. Requests for permission or further information should be addressed to the Permissions Department, John Wiley & Sons, Inc.

Library of Congress Cataloging in Publication Data:

Dougherty, E. M. (Ed M.)
    Human reliability analysis: a systems engineering approach with nuclear power plant applications/E. M. Dougherty, Jr. and J. R. Fragola.

    p. cm.
ISBN 0-471-60614-6.

    1. Nuclear power plants—Human factors. 2. Nuclear power plants—Reliability. 3. Systems engineering. I. Fragola, Joseph R. II. Title.

TK9153.D68 1987
621.48'35—dc19
                                                                87-31652
                                                                       CIP

Printed in the United States of America

10 9 8 7 6 5 4 3 2 1

Contents

List of Figures *ix*
List of Tables *xi*
Preface *xiii*

PART 1   *Fundamentals*

Chapter 1   **Introduction**   1
Chapter 2   **Historical Background**   3
Chapter 3   **Theoretical Issues**   11
   Toward a Theory of Human Reliability   13
   A Conceptual Framework   18
   Implications for Human Reliability Technology   20
Chapter 4   **HRA as Reliability Analysis**   25
   The Concept of Reliability   25
   Time-Dependent Human Failure Models   27
   Time-Independent Human Failure Models   34
Chapter 5   **HRA Data**   39
   Development of Existing Data Sets   40
   Data in THERP   47
   Data from Nuclear Plant Simulators   48
Chapter 6   **THERP**   59
   Task Analysis as a Human Reliability Model   59
   The HRA Event Tree   65
   Extending THERP   67

PART 2   *An Integrated Approach*

Chapter 7   **HRA as Part of Risk Analysis**   71
   Overview of PRA   74
   Human Failures and Accidents   76
   Incorporating HRA into a PRA   77
Chapter 8   **The HRA Procedure**   79
   An Overview   79
   Integration Tasks   79
   Initial Quantification Support Tasks   83
   Detailed Event Analysis Tasks   83

Contents, Cont.

Chapter 9   **Qualitative Analysis**   **85**
Plant Familiarization   85
Classification of Human Failures   86
Guidelines to Systems Analysts   90
Identifying Reliability Influences   98

Chapter 10   **A TRC System**   **107**
Basic Premise of a TRC Approach   107
Some TRCs Used in PRA   108
The Lognormal Family   112
Interpretation of the Data   116
Synopsis of the Patterns   123
Mathematical Formulation of the TRC System   124
Remaining Research Agenda   129

Chapter 11   **Quantitative Analysis**   **131**
The Approach   131
Screening   132
Quantifying Slips   133
Quantifying Mistakes   138
ORCA   142

PART 3   Topics and Conclusions

Chapter 12   **Burden**   **149**
Sources of Burden   150
A Qualitative Example   153
Sliding Time Windows   156
Potential Quantitative Solutions   157

Chapter 13   **Garoña Recovery Analysis: A BWR Example**   **165**
The Transient Sequences and Human Reliability Questions   166
TW — An Old Concern   168
Different Conditions, Same Action   169
Event Interpretation   169
A Photo Survey   173
Quantification   178
Conclusions   179

Chapter 14   **Feed and Bleed: A PWR Example**   **181**

Chapter 15   **Conclusions**   **185**
Limitations   186
Future Directions   187

Contents, Cont.

Appendix	**Introduction to Probabilistic Risk Assessment for Nuclear Power Generating Stations** **189**

                 Varieties of PRA   189
                 The Level 1 PRA   190
                 Organization of a Level 1 PRA   191

Acronyms   **203**

Bibliography   **205**

Glossary   **217**

Index of Authors   **223**

Index of Topics   **229**

List of Figures

2-1	Rasmussen's model	7
2-2	Embrey's taxonomic system	8
2-3	An information processing view of a nuclear power plant	9
3-1	The world within	19
4-1	Plot of response data on probability paper	29
4-2	Plot of data on log-probability paper	32
4-3	Density for the data in Table 4-1	35
4-4	Distribution function of the data in Table 4-1	35
4-5	Solution rate of data in Table 4-1	35
4-6	The time reliability correlation from the data in Table 4-1	36
5-1	The trail from raw data to information	39
5-2	Some of the ORNL simulator data	51
5-3	An aggregation of the EdF simulator data	52
5-4	Vigilance response data	52
5-5	TRCs for generic action types from the data from the LaSalle simulator	55
5-6	Data from LaSalle simulator for selected risk-significant specific actions	57
6-1	Major activities and potential failures in the task of mitigating a small LOCA	62
6-2	The overall THERP approach to performing an HRA	64
6-3	An HRA event tree model of the task of mitigating a small LOCA	65
7-1	The PRA process	72
7-2	The levels at which people interact with process plants	73
7-3	Context of human failures and errors	76
8-1	The HRA process	80
9-1	Classification system from human failure events	91
9-2	Level of incorporation of human failure events in PRA models	92
9-3	Fault tree model of failure of recirculation ECCS	96

List of Figures, cont.

9-4	Recovery fault tree model for recirculation failure event E 98	
9-5	Types of influences and their scaling 102	
10-1	TRCs based on judgment used in some PRAs 108	
10-2	A typical TRC with two percentile TRCs to represent uncertainty 115	
10-3	Response time data that formed initial basis for TRCs 118	
10-4	Solution rates for some of the LaSalle TRCs 122	
10-5	TRC system curves relative to the simulator data ranges 129	
11-1	A sample operator action event tree 132	
11-2	The screening strategy 133	
11-3	Process for estimating the probability of slips 135	
11-4	The basic time reliability correlations 141	
11-5	ORCA logic diagram 146	
12-1	Sliding windows indicating burdened times 156	
12-2	Decrement in reliability tasks 158	
12-3	Convolving a TRC, 1, 2, and 3 times 161	
12-4	The available time required to achieve a threshold probability of failure 161	
12-5	Correlation of conflict to convolution 163	
13-1	Transient event tree for Garoña plant 166	
13-2	Operator recognizes transient type 175	
13-3	Operator attempts to obtain IC flow 176	
13-4	Lack of HPCI flow is noted 176	
13-5	Operator attempts to use low pressure systems 177	
14-1	The June 1985 loss of feedwater at Davis-Besse 183	
A-1	The interrelationship of tasks in a level 1 PRA 193	
A-2	The top-down approach to accident sequence identification 195	
A-3	The modularization process in systems modeling 197	

List of Tables

2-1	Berliner classification of tasks	4
2-2	Role of intrinsic and extrinsic factors in influencing errors	8
3-1	A component failure taxonomy	11
3-2	A human error taxonomy	12
3-3	The influences of human performance on human reliability theory	21
4-1	Simulator response time data and statistics	28
4-2	Relating lognormal to normal funtional forms	30
4-3	Relating lognormal parameters to response time parameters	31
4-4	Discrete and calculated non-response statistics	34
4-5	Comparing time-dependent reliability for human and hardware failures	36
4-6	Continuing the human and hardware analogy	37
5-1	PRAs included in BNL HRA project	44
5-2	Some typical values from THERP	49
5-3	Recovery actions aggregated into generic action types LaSalle Simulator Exercises	54
5-4	TRC parameters for LaSalle generic actions	56
5-5	TRC parameters for specific LaSalle actions	57
6-1	Example task analysis	60
6-2	THERP dependency model	63
6-3	Manning model in THERP	66
9-1	HRA information needed from system analysts	95
9-2	Original cutsets for recirculation failure	97
9-3	Human failure event decomposition	99
9-4	Significant influences in prototypical situations	100
9-5	Significant influences on human reliability	101
9-6	SLI calculus	103
9-7	SLIM calculation for recirculation event, $E_{mistake}$	104
10-1	THERP manning model	109
10-2	Lognormalizing the THERP diagnosis TRC	114
10-3	ORNL median times	117
10-4	ORNL error factors	117

List of Tables, cont.

10-5	LaSalle median times	120
10-6	LaSalle error factors	121
10-7	Parameterizing the TRC range	124
10-8	Time reliability correlation values, rule-based, without hesitancy <2, 3.2>	127
10-9	Time reliability correlation values, rule-based, with hesitancy <2, 6.4>	128
10-10	Time reliability correlation values, recovery, without hesitancy <4, 3.2>	128
10-11	Time reliability correlation values, recovery, with hesitancy <4, 6.4>	128
11-1	Fine screening procedure for latent failures	138
11-2	Quantification procedure for mistakes	143
11-3	ORCA Human failure event record sheet, mistake	144
11-4	ORCA Human failure event record sheet, slip	145
12-1	Sources of operator burden	153
12-2	Effects of multiple convolutions of the base TRC	160
12-3	Available times (min) to achieve various failure thresholds	162
12-4	Correlation of convolution to conflict	12-4
13-1	Event interpretation	171
13-2	Failure modes of KLO	172
13-3	Operator actions in sequences TKCNW	174
14-1	A SLI calculation	184

PREFACE

The work that led to the development of this book and to the development and synthesis of the human reliability techniques presented was conducted over a period of eight years. During that time, the authors were associated with other organizations. However, the critical period in the evolution of thought and practices occurred while the authors were engaged in the performance of nearly a dozen probabilistic risk assessments (PRAs) of a variety of nuclear facilities while employed in the Advanced Technology Division of Science Applications International Corporation (SAIC).

The motivation for this work began when the authors recognized the problems in treating human failures in the Reactor Safety Study (USNRC, 1975), the first nuclear power plant PRA. These problems had been highlighted by the criticism of the RSS by the Lewis Report (Lewis, 1976). Our involvement in this area occurred soon thereafter when the pioneering efforts of Swain, which led to the publication of his THERP approach (Swain, 1983), did not seem to properly address what we considered to be the more risk significant human failures. The accident at Three Mile Island and the followup Kemeny (1979) and Rogovin (1979) reports seemed to confirm our convictions. Our interaction with colleagues at the two Myrtle Beach Conferences (Schmall, 1980 and *Conference Record*, 1982) convinced us that our concerns were real. Discussions, particularly with Jens Rasmussen and John Wreathall, helped to focus our attention toward the development of a systematic approach which is recorded here.

This book began as a set of papers and reports which we published separately and then combined into an internal SAIC report which received wide distribution. The SAIC report was expanded several times as our methods were refined and we saw the need to add additional materials from our performance of Human Reliability Analyses (HRAs) in support of various PRAs and from developing course material for our HRA training course. The report eventually became substantial enough and the demand for it wide enough that we felt it was appropriate for the work to be published in a more permanent and more readily available medium.

It is not our claim that this book captures all the relevant research in the area of HRA, but we have attempted to provide a comprehensive set of references to the body of work that we know exists. It is further not our claim that this book will represent the end to developments in the human reliability discipline; quite the contrary, HRA is a fledgling enterprise. The recency of some of the references indicates how this young field is rapidly developing. What we do claim is that this book presents one relatively comprehensive approach to performing HRA, which has been attempted to be developed within the framework of the systems engineering tradition. Readers of this book, although they will not acquire all of the skills necessary to perform a full HRA (because the final plant and sequence-specific quantification is still very much an art rather than a science), will acquire enough understanding of how an HRA is performed, how to incorporate human failures into PRA models, and how to systematically screen human failures for their risk significance.

As the title implies, the thrust and the examples in this book are taken from the nuclear power industry, where our primary experience lay. However, we feel, from our experiences with the offshore oil, chemical process, and aerospace industries, that the approach has broader applications. We have limited the title only so as not to disappoint the reader who might be interested in specific examples taken from other industries.

Finally, our intention in writing this book was to inform the readers as to what human reliability analysis is and how it is applied. We hope that this work accomplishes that goal.

People of the Advanced Technology Division of Science Applications International Corporation have played an instrumental role in advising, reviewing, and applying the methods herein. In particular, Martin Stutzke developed ORCA and refined the statistical techniques of time reliability correlations, Erin Collins spearheaded the Garoña analysis and provided much needed editorial services, and Karen Troy helped in the technical review and mechanics of producing the book.

PRA and HRA are international enterprises and this work owes much to the free play on the ideas of Barbara Bell, David Embrey, Paul Haas, Bill Hannaman, Jens Rasmussen, Jim Reason, Bill Rouse, Tony Spurgin, Alan Swain, David Worledge, and John Wreathall. No intent, however, is meant that the ideas that made it into the book are in any way representative of the ideas of the aforementioned. The facts may be due to them; the errors are the authors'.

Ed M. Dougherty, Jr.   Joseph R. Fragola   1987

PART 1
Fundamentals

Chapter 1
INTRODUCTION

Today's technologies have inherent risks that affect and are affected by the actions of people in normal operation and maintenance of the technologies and, of course, in emergency operation. The nuclear power plant industry has estimated risk due to human action or inaction at 50 to 70%; offshore petroleum drilling hazards apparently arise in some 70% of the cases from human causes; pilot judgment is attributed to at least 50% of the fatalities from commercial air travel (Jensen, 1982).

These statistics simplistically assume that just because a user of a technology "had it last" that any contribution to its failure from his or her actions necessarily meant that the failure was his or her fault. In reality, failure begins with the design of and decision to build a technological system. Errors in design and construction are numerous and are often misinterpreted as user failures. But more to the point, the type of system built—its complexity and hazards—potentially places the user in situations in which a high likelihood of success cannot be expected **as the system is designed**. Accidents are "normal" (but not necessarily frequent) attributes of complex and highly interactive systems, such as a nuclear power plant (Perrow, 1984). The errors of operators in such technologies are often forced by the system and its circumstances; and in any case, error is always the product of the same behavioral mechanisms that make success a possibility as well (Reason, 1987a).

From this perspective, it may be concluded that risk will always have a human factor. This human contribution to risk can be understood, assessed, and roughly quantified using techniques of human reliability analysis (HRA). Human reliability is defined analogously to that for hardware as the probability that some set of human actions is performed successfully in a specified time or opportunity. Several HRA techniques have been integrated into a systematic approach for use in conjunction with probabilistic risk assessment (PRA), control room and operator aids design review, human factors analysis, plant availability assessment, and other special reliability and risk studies. The HRA approach described in subsequent chapters has been used in various evolutions in the PRAs of the following nuclear power facilities:

Catawba	Clinch River Breeder Reactor
Crystal River-3	Davis-Besse
Garoña	Hanford N-Reactor
McGuire, Unit 1	Oconee, Unit 3
Savannah River Laboratory Reactor	Shearon Harris, and
	Shoreham.

In every study, the HRA led to findings that potentially influenced the risk profile for the facility and to recommendations the measurably reduced that risk.

That HRA technology is timely is indicated by the number of recent books published that address HRA in some way (Park, 1987; Dhillon, 1986; Goodstein, et al., [1988]; Rasmussen, et al., 1987; Shooman, [1987]). None provided an approach to HRA integral to PRA.

In attempting to provide that integration, Part 1 of the book describes the history, concepts, and fundamentals of HRA, ending with a description of the first comprehensive approach to HRA used in nuclear plant PRAs. Part 2 presents the integrated approach developed over the course of performing the above-mentioned PRAs. Part 3 provides an introduction to a new research topic in HRA—burden—and provides two examples of HRA-like analyses and then concludes. An appendix describing one way to perform a PRA is provided at the end of the book, along with a bibliography, a glossary of terms and acronyms used throughout the book, and an authors index and a topical index.

Chapter 2
HISTORICAL BACKGROUND

Human reliability analysis is an offshoot of the analysis of human performance in an industrial setting, which in turn, is one of the many human factors concerns in industry. The roots of HRA can probably be traced back to the beginnings of the Industrial Revolution but HRA began to evolve into a discipline in the 1950's. Under the auspices of industrial engineering and behavioral psychology, initial efforts were made to investigate the influences of the human being in the performance of tasks (Maynard, et al., 1948).

The development of the field of human performance analysis was influenced by the first systematic treatments of the problem of reliability in complex systems. The publication of early texts authored by Bazovsky (1961), Lloyd and Lipow (1961), and the research of Birnbaum (1961) in the United States was paralleled in the Soviet Union by Gedenko (1969). These works established the theoretical setting for the new field. With the later publications by Sandler (1963) and the more comprehensive works of Barlow and Prochan (1965) and Shooman (1969), reliability engineering was becoming a full-fledged discipline which combined the organizational tools of the systems engineering tradition with the theoretical tools of probabilistic analysis. The successful application of this new engineering discipline to the evaluation of hardware systems motivated practitioners to attempt to apply the same tools to the evaluation of human-machine systems.

However at the time, there was very little in the way of human factors data and also no accepted human performance theories or models. The realization of this led to a research project which produced a workable collection of human reliability figures known as the AIR (American Institute for Research) Data Store in 1962 (Munger, et al., 1962). In 1964, several approaches to quantifying human performance were developed using the AIR Data Store (*Human Factors*, 1964). A noticeably missing factor in these quantification schemes was a systematic approach to the classification of human performance in various tasks. Classification structures based on behavior were attempted by several people, among them Berliner (Berliner, 1964).

The Berliner development allowed for tasks to be decomposed into elements from a behavioral perspective (see Table 2-1, which is reproduced from Woodson, 1981), with the intention that this, in turn, would enable a more accurate application of data (what little was available at the time) to specific human actions. These data schemes begat the successful and consistent application of "task analysis" to human performance analyses but did not induce improvements in data collection. In this same time frame, the human performance model problem was being studied by

Table 2-1
Berliner Classification of Tasks

Processes	Activities	Specific Behaviors
Perceptual	Searching for and receiving information	detect inspect observe read receive scan survey
	Identifying objects, actions, and events	discriminate identify locate
Mediational	Information processing	calculate categorize compute encode interpolate itemize tabulate translate
	Problem solving and decision making	analyze calculate choose compare compute estimate predict plan
Communications		advise answer communicate direct indicate inform instruct request transmit
Motor	Simple, discrete tasks	activate close connect disconnect hold join lower move press set raise
	Complex, continuous tasks	align regulate synchronize track transport

With permission from W. E. Woodson, *Human Factors Handbook*, © 1981, Mc-Graw Hill Book Co., adapted from C. Berliner, et al., "Behaviors, Measures and Instruments for Performance Evaluation in Simulated Environments", *Symposium and Workshop on Quantification of Human Performance at Albuquerque, NM*, 1964.

Swain and his collaborators. His early work (Swain, 1964) was later refined into the well-known Technique for Human Error Rate Prediction (THERP).

In the late 1960s, analysts recognized the need to account for the situational influences on task actions as well as the elements of human behavior. Altman (Askren, 1967), in particular, described the combination of situational and behavioral influences in terms of a molar, or data cell classification scheme. This data cell concept was enhanced (Meister, 1969) with a description of equipment operation functions in terms of "task units" comprised of the equipment operated and the action of operation. Identifying the appropriate units to be quantified using reliability techniques is still an issue. Additional data banks arose, but most consisted of data reliant upon human judgment to determine the relative influence of task parameters on human reliability. The Sandia Human Error Rate Bank (Rigby, 1967), however, was created with the intent of using actual human performance data to be collected and codified on a continuing basis. As the 1970s began, much of the new work in human reliability data and analysis field was initiated by studies funded by the military. The Navy held seminars on human reliability in 1970 (Jenkins, 1970) featuring presentations by A. Seigel and A. Swain, both of whom contributed heavily to HRA development during this time. Design of systems and equipment accounted for human body specifications using three source books resulting from military studies: Van Cott and Kincade's Human Engineering Guide (Van Cott and Kincade, 1972), the Bioastronautics Data Book (Parker and West, 1973), and the MIL Standard of Human Engineering Design Criteria (US DoD, 1970).

The 1975 Reliability and Maintainability Symposium (Proceedings, 1975) exhibited examples of HRA models which had been extended to different applications earlier in the 1970's. These proceedings included a paper on the Siegel, Wolf, and Lautman efforts to produce "a set of stochastic, digital simulation models for simulating the performance of the human component in man-machine systems..." In the same proceedings, Swain and Guttmann described the application of THERP to the nuclear power plant environment. This paper was a brief overview of the work that they had performed for the Reactor Safety Study, WASH-1400 (USNRC, 1975). Estimates of human reliability were produced using basic estimates of human task performance during normal and high stress situations modified by the environs and other "performance shaping factors". Also in this time frame, the US Navy published a user's manual for their NAVSEA Human Reliability Prediction System (US DoD, 1977). This manual not only described the Navy's approach in quantifying human errors in electronics systems operations, but included calculational performance predictions applied to test, maintenance, and personnel selection situations. This approach could be encoded to produce estimates of error or success probabilities and error rates using data obtained from in situ observations, historical records, or simulators.

Then, an event occurred in 1979 which greatly influenced the perspective of the nuclear community concerning human reliability in nuclear plants. In March, the accident at Three Mile Island (TMI) occurred, placing nuclear power into an unflattering spotlight, yet forcing the industry to recognize much more directly human fallibility. The report resulting from the President's Commission (Kemeny, 1979) charged with review and assessment of the accident, found that "inappropriate operator action" resulting from training and procedural deficiencies, failure to learn from previous incidents, and deficient control room design caused the TMI accident. [Note that some people, including Perrow (1984), have noted the unfairness in attributing the operators with any such blame.]

A workshop was convened that December to discuss human factors and nuclear safety. This first "Myrtle Beach Conference" brought together representatives from engineering, psychology, reactor operation, and HRA. One of the priority research areas identified from Myrtle Beach I was for a "systematic, consistent, and reproducible approach for the quantitative evaluation of the reliability of the human component in the system" (Schmall, 1980). An initial draft of the documentation of THERP was available at the conference, which included one such approach to HRA.

A solution to the human performance classification problem resulted from a "model" created by Jens Rasmussen and published in RISØ-M-2240 (Rasmussen, et al., 1981). Figure 2-1 shows the model. The model was a compilation of ideas from the information and cognitive sciences that represented the mechanisms a person uses in going from noticing an indication of an off-normal event to acting on it. This model included sufficient cognitive steps to lend insight into the kinds and causes of errors across the spectrum of process plant tasks.

The Rasmussen taxonomy followed the Berliner classification in that it began with a perceptual process, split into detection, observation, and identification, and ended with a motor process he called execution. Communication was not explicitly modeled. However, the mediational process was split into interpretation, evaluation, task definition, and procedure formulation. Several "shortcuts" to the complete trail from detection to execution were recognized. These included a "reflex", or release of a preset response, that connected the execution of the response directly to the detection of a need to respond. Another shortcut was a task recognition connection from identifying the system's state to choosing a well-known, specific procedure to handle the state. This model and its resultant taxonomy has become the de facto standard for analysis of human and system interactions in the nuclear industry.

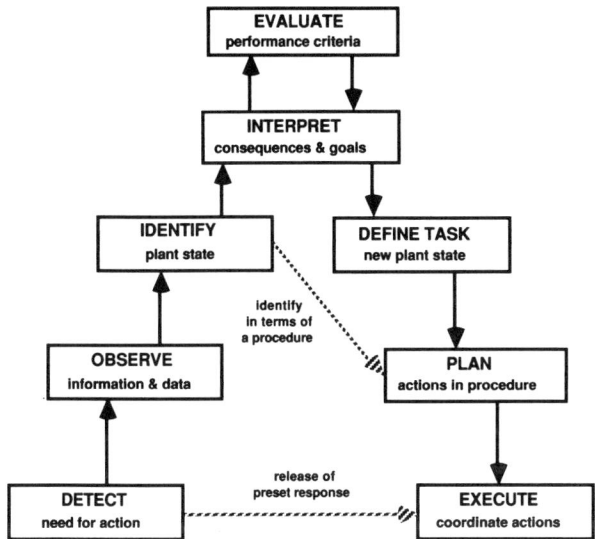

Figure 2-1. Rasmussen's model. (From Rasmussen, 1976.)

A recent variation is due to Embrey (Embrey, 1986). His system to classify operations tasks is depicted, slightly modified, in Figure 2-2. What Rasmussen called rule-based behavior is split according to whether there is a diagnostic element, as in fault management during off-normal incidents. If diagnosis or decision making is needed but no rules are available to assist the activity, then operators must act based on deeper, more fundamental knowledge. Skills include pattern recognition and actions that are manual, well-trained, well-known, and practiced frequently. Otherwise, rule-based behavior is elicited. This taxonomy has been extended by Reason (1987b) to include a speculation as the role that intrinsic and extrinsic factors influence the types of errors that predominantly arise in the behavior classes. Table 2-2 shows the relative contributions of intrinsic and extrinsic factors to skill-based slips, rule-based mistakes, and knowledge-based mistakes. Intrinsic factors include cognitive biases, attentional limitations, and limitations to human rationality. Extrinsic factors include task characteristics, effects of the situation, and factors from the environment.

A second Myrtle Beach conference was held in September 1981. A working group was specifically set up to examine the state-of-the-art in human performance analysis. Swain's THERP models, a cognitive-oriented model of Moray, the Operator Action Tree (OATS) model of Wreathall, and an HRA approach of Hannaman were reviewed (*Conference Record*, 1982). The "ideal" model, it was agreed, would combine the features of all the types, by basically enhancing the use

8 Human Reliability Analysis

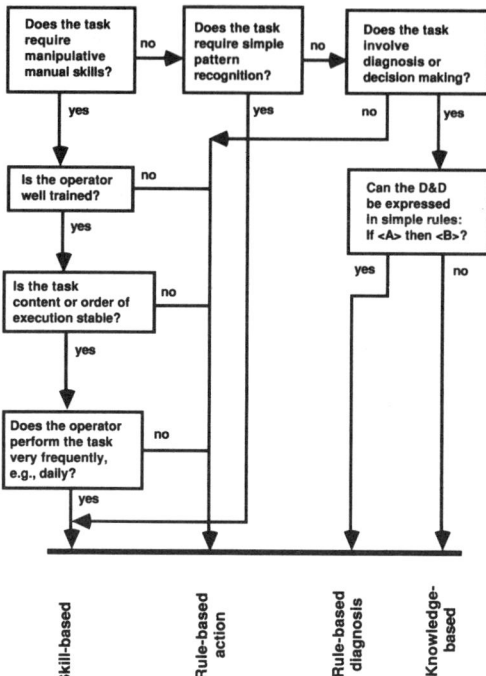

Figure 2-2. Embrey's taxonomic system.

of a cognitive model with a top-down systems engineering approach. This model should be verifiable, meaning that the process used to derive the human reliability estimates could be reproduced, thus showing the model to be internally consistent. In addition, the needs of users from both the systems analysis and the human factors analysis perspectives needed to be addressed by providing access to information from either an equipment or behavioral standpoint. At that time, it was agreed that THERP provided an acceptable model for certain situations but that some cognitive modeling was necessary to provide insight into operator behavior, especially in off-

Table 2-2
Role of Intrinsic and Extrinsic Factors in Influencing Errors

(Used with permission from J. Rasmussen, et al.,
New Technology and Human Error, © 1987 by John Wiley & Sons, Ltd.)

	INTRINSIC	EXTRINSIC
Skill-based Slips	high	moderate
Rule-based Mistakes	moderate	high
Knowledge-based Mistakes	low	very high

normal process conditions. The OATS model was recommended as an interim bridge from THERP to the necessary cognitive-oriented model and an effort was undertaken to provide a systematic data set for using THERP in a more procedural setting (Fragola and Chang, 1983). Notably, however, the *PRA Procedures Guide* (USNRC, 1982) adopted THERP as its only recommended HRA approach in 1981.

Proponents of the need to model cognitive influences on human performance saw the process man-machine interface as a complex information system, as depicted in Figure 2-3 (Fragola and Chang, 1983). A time reliability correlation (TRC) concept was adopted to model the needed time to pass and process information. This amounts to a "model" of the thinking time effects on human reliability, a strategy used as early as 1969 in Britain (Ablitt, 1969 and Greene, 1969) and later in the U.S (Fleming, et al., 1979 and Fullwood and Husseiny, 1979). The NRC-sponsored Interim Reliability Evaluation Program developed a correlation to be used in accounting for operator recovery from failed equipment (Kolb, et al., 1982) and the industry also began to use TRCs to model recovery (LILCO, 1983). Efforts of Fragola and those of Wreathall in the development of OATS, compiled

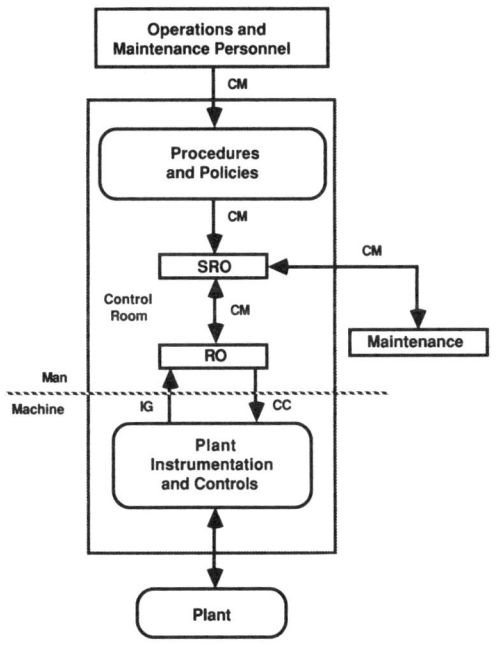

Figure 2-3. An information processing view of a nuclear power plant.

research on the time reliability correlations. A synopsis of this work appeared in NUREG/CR-3010 in November 1982 (Hall, Fragola, and Wreathall, 1982). Swain, too, adopted a time reliability correlation to be applied to the initial diagnosis needed to choose the correct procedure at the beginning of an off-normal event (Swain and Guttmann, 1983). This "nominal diagnosis model" was documented in Chapter 12 of the final version of the THERP handbook in August 1983. The approach presented in this volume uses time reliability correlations extensively and was developed in conjunction with the PRAs of Oconee-3, McGuire-1, the Clinch River Breeder Reactor, Catawba-1, and Crystal River-3 plants. During this effort to develop classification and quantification techniques, an often overlooked ingredient in HRA was addressed seriously for the first time. In the nuclear climate since the accident at Three Mile Island in March 1979, most human reliability work was undertaken in support of a plant's probabilistic risk assessment. It was not self-evident, however, just how the HRA should be integrated within the PRA effort. WASH-1400 had built system failure models down to the component level and then included generic events to account for human-induced failure modes. In this approach, the HRA was considered an independent adjunct to the PRA (not unlike other parts of the PRA).

However, TMI and other significant "precursors" to accidents showed that actions by operators can propagate effects throughout an accident's sequence of events. Thus, HRA considerations need to be factored into PRA at a much "higher" level than that of equipment components. The Oconee PRA initiated the nuclear utility industry efforts in integrating HRA and PRA. The Electric Power Research Institute later funded a program to summarize these efforts being developed ad hoc in all subsequent PRAs. The result was an approach called the Systematic Human Action Reliability Procedure (SHARP - Hannaman, et al., 1984), released in interim form in June 1984. The integration issue continues to be investigated (IEEE, 1987).

Human reliability analysis is still a research and development activity. However, the last twenty years or so have resulted in HRA evolving into a credible and useful discipline that can be applied to risk analyses, hazard analyses, experience database development, and the development of computer-based systems that supplement or substitute for human expertise. In fact, as automation of process control continues, new problems relative to human errors arise (Weiner and Curry, 1980). With the advent of so-called expert systems, there seems to be a trend towards automating decision making, thereby taking human expertise out of the control room altogether. The assets and liabilities of this new wave of human-machine systems are only beginning to be recognized (Dreyfus and Dreyfus, 1986).

Chapter 3
THEORETICAL ISSUES

There are several technical problems in trying to assess human reliability in a risk setting and because HRA is a controversial, fairly new discipline, each problem is at issue in the HRA community (Adams, 1982, for example). The dominant issue is that there is no theory of human behavior that is both well-accepted in the various human sciences and complete enough to use to develop a theory of human reliability. However, human reliability analysis is founded on the idea "that the seemingly immense variety of human errors observed may reflect complexity of the environment, rather than complexity in the psychological mechanisms involved" (Rasmussen, Duncan, and Leplat, 1987). It should be noted also that "predictable error and correct performance are two sides of the same coin, and hence demand a common explanatory principle" (Reason, 1987a).

Hardware system reliability theories, on the other hand, are based on observational data that is robust enough to support statistical models and rich enough to show patterns in failure types and causes. Table 3-1 provides a component failure taxonomy for some typical equipment.

Table 3-1
A Component Failure Taxonomy

| Component Types | Failure | | |
	Modes	Mechanisms	Causes
batteries	fail to operate	motor failures	too high
circuit breakers	fail to start	bearing failure	temperature
pumps	fail to run	seal failure	lack of lubricant
valves	fail to change	contact failure	oxidation
etc.	state	command fault	misalignment
	spurious operation	control instrument failure	contamination

In an analogous fashion, a human behavior/failure categorization was developed by Jens Rasmussen of Denmark's RISØ National Laboratory, which was based on a stimulus-cognition-response model (see Chapter 2) that traced the process people use as they act in response to an indication of something wrong. The result was a taxonomy (Rasmussen, et al., 1981) that seems to be rich enough to identify error modes and many of the error mechanisms in process task

11

performance, including emergency operation. Table 3-2 draws from the Rasmussen taxonomy and other ideas to depict a scheme for classifying human errors that includes two error types, several modes (based on the Rasmussen model), many mechanisms of error (based on Reason [1987a]), and many proximal (nearest) causes of error.

Table 3-2
A Human Error Taxonomy

Behavior Types	Error		
	Modes	Mechanisms	Causes
mistakes slips	misdetection misdiagnosis faulty decision faulty planning faulty actions	false sensations attentional failures memory lapses inaccurate recall misperceptions faulty judgments faulty inferences unintended actions	misleading indicator lack of knowledge uncertainty time stress distraction physical incapacitation excessive force human variability

The taxonomy is only suggestive. A theory of human reliability can only come as more is known about the human being, its source — particularly, human behavior, human capacities and functions, and the performance that results from these human factors. Such a theory of human reliability must arise from the human sciences, such as psychology, neurophysiology, cognitive science, and the brain sciences. Bits and pieces of facts and theory exist but are scattered over the collection of human sciences.

Part of the reason that no full theory exists yet is that people are the most complex entities around. They "obey the laws of physics ... [but] are designed to be so sensitive to the passing show that they never can be in the same microstate twice" (Dennett, 1984). Complicating this is the fact that the environment dealt with on a daily basis, even more so during incidents of stress and uncertainty, is so complex as to defy any foreseeable procedure to abstract and simplify into useful models, the primary tools of human performance theory. As organisms, people routinely deal with immense complexity using immense complexity. It may be that any simple model of the human mind is doomed to be but a simple-minded model.

Toward a Theory of Human Reliability

Surprisingly little is known about people and why they do what they do. Psychology is a century old as a science and is yet splintered into dozens of "schools", from behaviorists to existentialists to cognitivists to psychoanalysts. The "ideal" situation for HRA would consist of a validated theory of behavior that could be translated into a model of human/system performance that then could be quantified probabilistically. An alternative would be for enough data to detect theoretically meaningful patterns in human failures and to quantify the patterns with statistical robustness. Neither alternative exists at this moment.

However, the last few decades have revealed a lot about human behavior, albeit splintered and inchoate. Several areas of human performance are known to need accommodation in order to progress toward a theory of human reliability.

The nature of the mind/brain

The history of inquiry into the source of human behavior carries with it a riddle—the so-called mind/brain(body) dichotomy. The mind has always been elusive; a source of mysticism, religious contention, and philosophical speculation. It is often thought to be the essential characteristic of that which is human. The brain, on the other hand, is clearly the locus of activity that must in some way be correlated with most, if not all, human behavior.

The mind may "just" be the brain by another name—people have a natural affinity to objectify even abstract concepts and name them (Macnamara, 1982). Alternatively, the mind may be some part of the brain's activity, or may be but a matter of a difference in point of view—the mind being the brain's looking inward at itself. In any case, if the mind remains a mysterious, unscientific object, there is little hope for a foundational theory of human reliability.

Logic and architecture of behavior

The brain is modular in some fashion, acting on many levels with the modules taking on an autonomy (Furst, 1979 and Gazzaniga, 1985) that resembles a homunculus or "little man within the man" (Dennett, 1978). These homunuculi evolved as more and more capability became strategically effective and are functionally self-contained and often independent of other brain activity. The result is that brain works on many levels, seeming to be a hetarchy of strange loops among these levels (Hofstatder, 1979). The architecture of the brain may force the architecture of human reliability models to be as complex as well.

It has been said that people are rational but this is only a capability, a possibility. There is no evidence that our brain is significantly constructed on the (formal) laws of logic. Human brains are analogical, not digital; the computer metaphor of the mind is notoriously weak (Dreyfus and Dreyfus, 1986). The brain's processor(s) and memory are not distinct; experiences are stored in some holonomic fashion (Pribram, 1976) that is not easily segregated into linear data space.

Much, if not most, of the brain's memory seems to be associated with concepts and ideas that are abstract, imprecise, and based on uncertain evidence. People are believers — they seek meaning and answers (Frankl, 1969), and routinely make theories to explain anomalous or partial information (Gazziniga, 1985). The human brain can also hold inconsistent beliefs readily, often doggedly (Hoffer, 1951 and Janis, 1982). This tenacity of belief and its influence on behavior can lead to disastrous results in accident conditions.

Relation to the world

The validity of the senses has been a controversial issue throughout history. Human performance in diagnosis and supervisory control may hinge strongly on the degree that sense data reflect real phenomena (Kelley, 1986). Even so, much, if not all, of people's perceptual relation with the world is a synergestic participation on their part. People see, to a great deal, what they anticipate they will see (Neisser, 1976). They report what is consistent with their world view of the moment, often irrespective of the facts (Janis, 1982). They believe what is comfortable or what is consistent with some or all of their context.

Much, if not most, human activity, including actions, can or do go on without self-consciousness of the activity. This not only means that people typically do not know how they do things but also means that they do not necessarily or reliably know why they do some things.

Consciousness is a self-known phenomenon that "emerged as an inevitable consequence of one particular evolutionary strategy" (Rose, 1976). At the cellular level, consciousness is probably a function of the neuronal number and the connectivity of the brain's neurons. However, consciousness is not a phenomenon that easily fits into psychological theories much less theories of human reliability.

Choice phenomena

In a fundamental sense, all human behavior follows from choice—the choice to act. People make decisions routinely. It is not evident, however, that people make decisions the way in which the decision theorists would have them do so (Tversky and Khaneman, 1974). People seek meaning (Frankl, 1969) and base their actions accordingly. The creative act (May, 1975), or flash of intuition, is largely an unexplained phenomenon that can have implications for a theory of human reliability.

A decision may be a non-random but an apriori unpredictable happening in the brain rather than the logical conclusion to some rational process (Nozick, 1981). Choice is a multifaceted phenomenon that is a product of our biological endowment (Dennett, 1984). At the conscious level, what seems to be free will may only be confirmation of an activity already in motion. What may be the physiological indicators of a choice, e.g., the P-300 EEG wave (Furst, 1979 and Lerner, 1987), lags the body's activation toward the chosen activity. Choice is so crucially a human behavior that it must in some way be accommodated in a theory of human reliability.

Behavior types

There are at least two, distinct categories of behavior. One is skills; the other has been referred to as knowledge-based behavior (Rasmussen, 1983) or planning behavior (Johannsen and Rouse, 1983). These behavior categories match the age-old distinction between "know-how" and "know-why". This distinction is also sometimes referred to as being between cognition and subcognition (Hofstatder, 1985).*

Skills are highly practiced activities that require little, if any, conscious thinking or monitoring. Skills are numerous. There are the skills associated with pattern recognition that is so prevalent in human perception, from faces to free form analogies (Bongard, 1970) to families of print type (Hofstatder, 1985). There are motor skills, such as walking and manipulating objects by hand. There are directly cognitive skills, such as referring (Macnamara, 1982) and anticipation (Neisser, 1976). Some skills seem to be inborn, such as referring or face recognition, and only

---

* This approach appears to be at odds with some of the practitioners in the cognitive sciences. In this field, all activity between the stimulus and its elicited response is considered cognitive. In HRA, it is agreed that all mental processes have cognitive content but some processes are "deeper" than others, thereby producing substantial differences in externally measured variables such as response time or error rates, which are of primary concern to human reliability analysts. This distinction is readily admitted by many cognitivists, even to the degree that it explains how some individuals (experts) can exhibit remarkably better external variables than others (novices) for the same cognitive task. For example, this distinction allows the human reliability analyst to account for the level of training and practice in quantifying the failure probability for groups with different levels of training when performing the same task.

their use needs to be learned (Macnamara, 1982). Other skills must be acquired, as are, for example, the skills of an operatic artist (Smith, 1985):

> "... The relationship of word to sound was crucial, she said. You couldn't think in either musical or verbal terms. You had to think in a way that combined both. Cherchez le fond, she'd say. Find the core. Then, she said, when you'd analyzed the score as completely as possible, you had to put the analysis behind you and let the role come alive. Come to life in your blood and bones, that's how she put it."

Knowledge-based behavior, on the other hand, involves higher, often conscious, cognitive activities, such as planning and decision making, and exercises the more abstract, sometimes propositional or rule-based knowledge structures. Knowledge-based behavior is often necessitated by uncertainty or novelty in the circumstances at hand. Knowledge-based behavior is seldom practiced (because it cannot be in a specific way) and is at root what most people mean by "thinking".

These behavior types seem to involve radically different mechanisms, which may mean radically different reliability characteristics. For example, human errors are similarly of two kinds (Norman, 1983). Mistakes are failings of cognition—inadequacies in planning, decision making, and diagnosis; whereas slips (or lapses) are failings in cognitive control—failures in the implementation and monitoring of actions.

The theory of slips is advanced over that of mistakes, probably because slips are more frequent and noticeable, while mistakes are less frequent and their signs are more complex and subtle. Reason (1983) has suggested that the cognitive control of action depends on a fundamental resource, attention. Since attention is finite, demands or workload can stretch this resource too thin. If the analogy of an electronic signal processing system is considered, the relationship between workload and error rate can be visualized more clearly as it relates to the human information processing system. If a signal processor is designed with a given bandwidth, then it will process signals with an acceptable error rate as long as the signal frequency is compatible with the bandwidth of the processor. Once the limits of the processor bandwidth begins to be approached, there is a marked increase in the error rate, until for a significantly high frequency, the processor breaks down completely. The extreme non-linearity of the relationship between error rate and input signal rate carries over to the human information processor. However, what does not carry over for the human system, at least in the case of trained, experienced individuals, is complete breakdown. What happens in the case of trained, experienced operators is that they limit the input signal rate by discrimination or increase the processing speed by accessing faster "rules of thumb" algorithms or both.

While this strategy allows the operator to continue to process input information, the result can be a failure in monitoring the activity, "place-losing" errors; or a poor distribution of attention, allowing errors of "strong habit intrusions". External influences induce stress that inhibits the efficient allocation of attention. In this view, human failures, i.e., slips, are only weakly correlated to the (potential) consequences of a failure. For this reason, accidents can be defined as errors "with sad consequences" (Cherns, 1962).

Mistakes, on the other hand, seem to involve a feedback process from the environment by means of the human capacities to anticipate and project (Neisser, 1980). Error likelihood can depend on the perceived consequences of the action (Johannsen and Rouse, 1983). Less is known about the mechanisms of the higher cognitive processes as they relate to errors. However, it is known that decision making in everyday settings induces a kind of trauma that makes the deciding stressful and often faulted (Janis and Mann, 1977). Psychologically, deciding produces "wonder and awe" or often a "dread or fear of failure", exposing a fundamental "vulnerability to anxiety" (May, 1981). As a result, errors that proceed from cognition seem often to result from influences not directly attributable to the processes themselves.

Emotive/valuative phenomena

People not only make decisions routinely but under the pressures of the moment. Hot cognition (Janis and Mann, 1977), not cold calculation, is the basis of human decision making; decisions are inherently traumatic (Janis and Mann, 1977) and existentially momentous (May, 1981).

People have a propensity to act under emotions or toward some value. As a result, stresses on human performance come from within as much or more as from the outside world. If the heat of the moment is an influence on human reliability, a theory of such must account for a myriad of forces, stressors, and influences.

World complexity

The objective driving force behind human behavior is the world at hand. This world, especially a high-technology environment, is a complex one which is "dynamic or event-driven that consists of many interacting parts, where evidence is uncertain and where actions involve risky outcomes" (Woods, 1986). The uncertainty of the world and its events provides an objective impetus for the stochastic basis of reliability in a theory of human reliability.

Social Interactions

People thinking and acting together do not necessarily do so harmoniously. Their behavior may take on the same error phenomena as a group that individuals exhibit; for example, one or more team members may defer to a "leader" with disastrous consequences (Trivers and Newton, 1982). Group behavior may even become irrational, ignoring and suppressing evidence against desired beliefs (Janis, 1982). The phenomena associated with social interaction as it relates to human failure may even invalidate the concept of human error in complex crew/hardware systems (Rasmussen, Duncan, Leplat, 1987).

A Conceptual Framework

Without the full ammunition needed for a theory, an alternative strategy used in theory construction is to generate a conceptual framework that loosely binds the isolated facts and characteristics that are known. Framework building is a popular enterprise. Two frameworks are particularly relevant to the issue of technological risks (Rasmussen, 1976 and Hess, 1987). An alternative version is depicted in Figure 3-1.

As the figure depicts, the human being consists of many modules that carry on selected activities. There are probably many mechanisms that recognize various aspects of the perceivable world, many mechanisms that control action, and there may be specialists as well in interpreting, planning, and choosing actions. To control this brainstorm of activity, there evolved some executive monitor and activity scheduler (Dennett, 1978). There is also reason to believe that not all of the executor function is a conscious activity (Gazzaniga, 1985), so that there is room for a higher, or at least coequal, conscious module.

The human relation to the world is through the senses directly but its input is quickly processed by recognizers, anticipators, and interpretors that manipulate information and direct the senses to seek new information. These information paths are a two-way process of receive and send; information is judged and judgment selects and ignores information. If the situation is novel or uncertain enough, this interpretative role may rise to the executor in the case of conflicting interpretations or to the conscious mind, to ponder on its implications.

Action is affected by means of the motor apparatus (hands, feet, etc.). Action may arise when a situation is so imminently recognizable or matches a common or habitual schemata (Neisser, 1976) that a choice is made base on past experiences and the action follows. Again, however, if the situation is novel or uncertain,

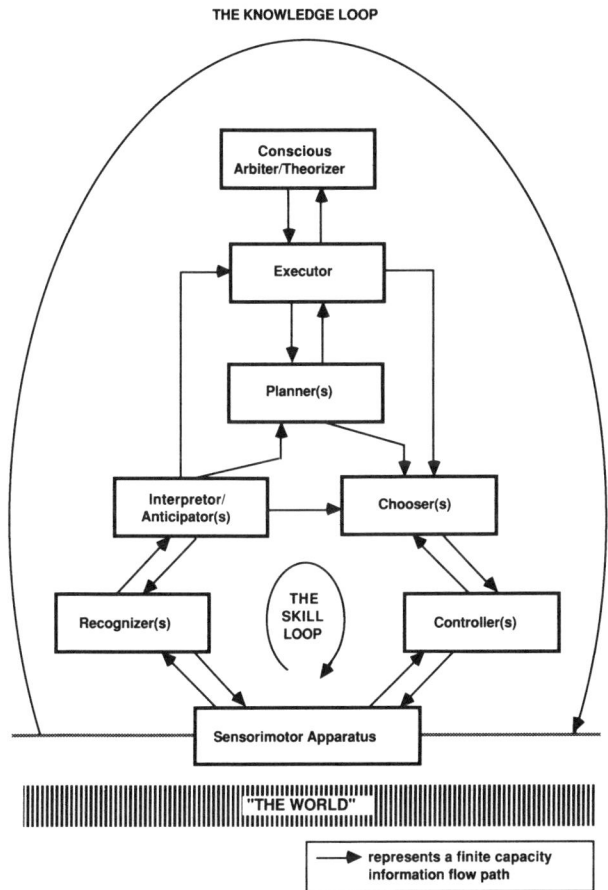

Figure 3-1. The world within.

significant planning may be needed, alternative plans may require executive or conscious arbitration and the choice may be the result of a full–fledged decision.

The behavior types are easy to recognize in this conceptual framework. Skills are deeply stored and apply to familiar, i.e., recognizable, routines, requiring little planning, execution, or conscious processing. The skill loop is shorter and presumably faster as an information flow from sensory input to action. On the other hand, the knowledge loop may pass through all categories of the modules, depending on the circumstances at hand. Consciousness has been likened to a limited channel transducer, and slows the process down considerably. Multiple interpretations of uncertain or novel situations can arise and force more extensive decision making. The mere fact of having to go beyond the single or packaged skill level strains the capacities of the person and results in burden (Hess, 1987).

The framework depicted in Figure 3-1 is a process diagram. There are many influences on human behavior that do not arise simply as one more module in the process. These influences may increase the effectiveness of a particular module, e.g., training may increase the "capacity" of the planning module(s) (Hess, 1987). Other influences may inhibit the performance of the module, e.g., uncertainty produces a physiological reaction called stress and slows down or paralyzes decision making ability.

Finally, the external world is not included in this framework. Others have felt obligated to model the universe (Woods, 1986a) along with human performance to it. This strategy is not justified for reasons of practicality perspective and does not seem necessary with our current state of knowledge on human response to world events.

Implications for Human Reliability Technology

The framework described above shows that a technology of human reliability, if not its theory, should account for the known dominant mechanisms of behavior and the extraneous influences on that behavior, whether they be from the environment or the world within. The theoretical areas of human performance identified earlier should have influence on the rudiments of a human reliability theory as summarized in Table 3-3.

Behavior Mechanisms

The basic categories of behavior are skills and knowledge-based behavior and human reliability technology must in some way account for both of them. Rasmussen (1983) also claims rule-based behavior as a third category. Rules are pretried sequences of skilled acts, often taking the form of condition-action associations. Rules may be explicitly inferred from experience or training and implemented in procedures. Rules are less highly practiced than skills, because they are less frequently used and more complex, and probably require extensive conscious monitoring, if not "thinking". However, it is not clear whether rules comprise a specific category of behavior or are merely a convenient way for us to report on the cognitive basis of our behavior.

Rules, however, seem to have significant reliability implications. Rules have been used in the nuclear power industry as a way to shortcut the full knowledge loop in a manner analogous to the skill loop. A rule, as the nuclear industry views it, is a simple situation-action pair, where the situation parameters are easily and directly discernible in the control room of a plant from instrumentation and the action is a

Table 3-3
The Influences of Human Performance on Human Reliability Theory.

Area of performance	Influence
The nature of the mind/brain	Defines the possibility of a first-principles theory
	Identifies the mechanisms of performance
Logic and architecture of behavior	Specifies cognitive content of a model
	Specifies the complexity of the model
Relation to world	Specifies factors to account for in diagnostic and supervisory control activities
Choice phenomena	Defines workload/burden
	Identifies crucial events to be modeled
Behavior types	Identifies the modes of failure to be modeled and the number of distinct models
Emotive/valuative phenomena	Identifies internal influences to be accounted
World complexity	Identifies external influences to be accounted
	Helps specify the events to be modeled
Social interactions	Requires manning model or other influence of crew behavior

simple set of control manipulations in the same vicinity of the control room. The simplicity of the rule, and its mandate as taught in training, are assumed to bypass the planner(s), executor, and conscious processor. Because of the human capacity to be aware of the implications of one's actions, such a rule can, however, actually increase operator burden by means of goal conflict (see Chapter 14 for an example) rather than make the intended behavior skill-like.

In hazardous technologies, high-risk events are induced by people while operating predominantly in the skill or rule regimes; handling the event, however, predominantly requires rules and knowledge. For example, maintenance is typically a know-how activity, whereas emergency diagnosis is a know-why activity. It is reasonable to believe that the reliability characteristics of know-how (skills and simple rules) and know-why behavior differ. Just how and how much is at issue in HRA. That it is a significant issue for risk analysis was demonstrated for the nuclear industry by TMI.

Influences on Human Reliability

Equipment performance characteristics are known to be influenced by environmental factors such as temperature. The U. S. Department of Defense has spent considerable resources in trying to understand and quantify these influences as reflected in the reliability-oriented MIL standards (e.g., US DoD, 1982) and handbooks (e.g., RAC, 1984). Human behavior is not only more wide ranging than equipment behavior but is, for that very reason, more susceptible to the environs, which includes the world within. The crew of the Titanic believed they were unsinkable and thus failed to take precautions that would have been normal on lesser vessels. An Air Florida pilot believed he could takeoff from Washington, DC's National Airport, even with indication that ice coated the wings, and seventy-eight people perished as the result of extreme self-deception (Trivers and Newton, 1982). Even minor things lead to accidents, as an Eastern Airlines L–1011 plane crashed into the Everglades while its whole crew was below the cockpit attending to an indicator malfunction (NTSB, 1973). Transportation is a process deemed so routine that we only think of it when we read grim headlines. Emergency conditions, as have occurred and will occur again in power plants and in other process facilities, may be much more susceptible.

One of the great influences on human cognition is its own limitations. Planning behavior in abnormal settings in piloting has been shown to vary with the perceived urgency of time and the occurrence of unanticipated events (Johannsen and Rouse, 1983). Planning is an effort and is a resource that must be allocated according to the circumstances. Studies of failure diagnosis show that diagnostic performance varies with time but not one-dimensionally. Maintenance troubleshooting seems to degrade as some degree of complexity is reached (Wohl, 1982). Time-to-diagnosis in experiments also relates to the number and kind of failure events and the strategy used by the diagnostician (Henneman and Rouse, 1986). Conflict between different goals creates hesitancy, a kind of cognitive lockup, and has been noted in nuclear power plant simulator exercises and in anecdotal reviews of actual events.

The mere fact of planning produces workload, which in turn, may use up available resources (Hess, 1987) and degrade the human performance. There is evidence that a physiological indicator of a decision is delayed as the burden on an operator increases (Lerner, 1987). And burden itself is a concept just now being explored as it relates to the reliability of operators in off-normal situations (Hess, 1987 and Black, 1987).

Theoretical evidence relating to the influences on cognition is beginning to accrue but from too many directions yet to synthesize into a coherent theory.

Probabilistics

Reliability is an intrinsically probabilistic concept. In hardware reliability theory, a potential event is specified, e.g., the successful performance of a piece of hardware over a period of time, or the successful operation of a piece of hardware on demand. Reliability is then the probability of the occurrence of that event. Since the event is concerned with success, or its complement failure, a specification of what constitutes success is required. Successful (and failed) performance arises out of the physical properties and mechanisms of the hardware, e.g., the growth of cracks in piping or the migration of electrons in semiconducting materials. Thus, a taxonomy of success or failure events should account for such mechanisms and properties at least at the level of observable effects, i.e., failure modes.

Analogously, a transformation from human behavior to human reliability theory must not only allow quantification to emerge but in particular the peculiar mathematics of probability and its interpretations. A prototypical manner in which probabilities are introduced in a seemingly deterministic process is to adopt the subjectivist stance—all knowledge, e.g., a theory, has uncertainty associated with it and this uncertainty obeys a calculus of probability. It is unlikely that the theory of human reliability will soon find any first-principles justification for its probabilistic requirements.

The issues of HRA are not just some academic exercise but represent an element in people's trying to manage, use, and survive their technology. It turns out that error is the price paid for the diversity and complexity that human behavior can exhibit. People are finite beings and error is an indication of as well as an effect of the finiteness of their resource management capability. The practitioners of HRA do not seek to eliminate human failure, an impossible goal, but seek to understand the consequences and causes of failures and to estimate the corresponding probabilities of occurrence. By doing so, the human reliability analysts can review human failures in a risk setting so that those of potential risk significance can be substantially reduced in the probability of occurrence or their consequence.

Chapter 4
HRA AS RELIABILITY ANALYSIS

Hardware reliability analysis consists of a mathematical theory, the techniques of data collection and its analysis, and several specific engineering methods. There are natural analogies between hardware and human reliability, although these analogies are limited, as may be inferred from Tables 3-1 and 3-2. This chapter discusses the mathematical and general data aspects of the analysis of human reliability. Subsequent chapters deal with data in detail and with the engineering-oriented methods of HRA. First, reliability is presented as a concept in a general enough form to include both hardware and human reliability.

The Concept of Reliability

Hardware reliability deals with failures over a mission. Some equipment is active in performing its mission, e.g., pumps, motors, and fans rotate in order to function. Other equipment is passive. Some passive equipment can or must change state in order to function, e.g., valves open and circuit breakers close to allow flow of water and electricity, respectively. Other passive equipment is "permanently passive", although its environment may change, e.g., piping and electrical cabling serve as material conduits and system boundaries but do not change state to perform. Equipment that is active can be active all of the time or can wait in standby. Equipment that can change state can be set in its desired functional state or left in another state. In these standby modes, equipment must be "demanded" to change state and/or activate. Thus, there are three failure regimes, depending on the type of equipment installed:

1. Failures while operating (active),
2. Failures while in standby, or
3. Failures on demand.

These failure types are referred to as "operating", "standby", and "demand" failures, respectively. However, the first two failure types occur over time and the last type occurs at a specific point of time; and this seems to be the distinguishing reliability characteristic.

Thus, a mission involves either some period of time for a performance or a point in time, associated with a demand to perform. Unreliability arises from the fact that equipment is always in contact with a potentially hostile environment, e.g., water, heat, mechanical strain, cosmic rays, etc., whether the equipment is active or passive. Failures occur over time as degradation processes evolve, such as crack growth in water-bearing components. These evolutionary processes eventually produce stresses on components or strains within the components that are large enough to surpass some capacity in the equipment's structural integrity.

On the other hand, some stresses occur suddenly; these are sometimes called shocks. Shocks, such as the vibration from a seismic event or the heat from a fire, may also exceed material capacities and cause failures. The mere fact of the activation of an active component stresses the component, so that demands to operate or change state are also shocks that may induce failure. Thus, although standby equipment may be subject to less severe environments on a normal basis, it then receives a shock each time it is activated.

These phenomena give rise to the measure of the rate of failures as occurrences per unit time or per demand. In recording statistics of failures to estimate such failure rates, the equipment is assumed to be initially in an operable, i.e., non-failed, state.

As a result, mathematically there are two types of reliability measures. Time-dependent failures can be assumed to evolve from a random variable—the time to first failure, **t**. In this case, reliability is expressed as:

$$R(t) = Pr\,[\mathbf{t} > t] \tag{1}$$

i.e., the probability that the first failure occurs after time, t.

For demand or shock dependent failures, the random variable is the "stress" of the shock, **s**. In this case, reliability is expressed as:

$$R = Pr\,[\mathbf{s} > \text{capacity of the object}]. \tag{2}$$

Human performance is also subject to environmental stresses and shocks. As Chapter 3 noted, the range and source of environmental influences are typically much greater for people than for equipment. The mathematics of the mechanisms of human failure is also slightly skewed from that of hardware. People can fail during "normal" activity. Also, a situation can "demand" human action to intervene or mitigate its consequences. Somewhat reversed from hardware, the normal human activity more often than not leads to time-independent human failures (slips) while the event demanded activity is more susceptible to time-dependent failures (mistakes).

A typical scenario for studying the reliability of nuclear power plant operators is in response to off–normal events. Some of these events can be simulated and the simulator has become a significant source of HRA data. As the event begins, the initial state of the person or crew of people who are to respond to the event is not successful (although not really failed, either). The performance measure is the time-to-respond adequately, or response time. As multiple crews are tested over the same scenario, these response times distribute over time due to various influences. It is then the number of responses relative to the total number of people or crews that can give rise to the probabilistic measures needed for reliability.

Other human failure modes or mechanisms do not appear to exhibit such a time dependency. It is likely that time-dependent mechanisms are going on in a person all the time but their time characteristics, e.g., process rate, do not always seem to dominate the reliability effects. A prototypical slip mechanism is task-capture. An example is a person who is driving to work but wants to stop at a convenience store on the way. There is a significant likelihood that he will drive straight to work and fail to stop at the store as intended. The reason for this is that the new task—drive to work while stopping on the way—is nearly identical to the old, habituated task of driving to work without the stop. Since the task is habituated, the person has not had to monitor the task frequently to assure its performance. If the person's monitoring frequency just happens to omit the topography around the store, then the old task may "capture" the new task and the stop will not be performed. It is difficult to find an explicit time relation in this kind of slip mechanism, as is the case for other slips. The probability of task capture would seem to more depend on the nearness of the new task to the old, the degree of habituation of the old task, the wherewithal of the person in initiating the task to provide some mnemonic device to alert himself of the imminence of the deviation from the old task, and other factors. For these reasons, slips seem more analogous to demand/shock failures in hardware than time-dependent failures, where the "demand" is the point of departure in the task-capture.

The next two sections discuss time-dependent and demand, or time-independent, human failure models.

Time-Dependent Human Failure Models

The risk from a process or transportation technology is clearly correlated to the response reliability of the people who operate the technology during unusual events. [That this risk contribution may be forced upon the operators and be beyond any reasonable expectation of human capabilities is an organizational problem not of the province of HRA.] The correlation of response times and human reliability will be developed in a manner similar to time-dependent models for hardware (Shooman, 1987).

Over the last decade, crew response time data has been accumulated from nuclear plant simulators for various events. A sample of data that can be inferred from the graphs included in the Oak Ridge National Laboratory (ORNL) simulator experiments (Bott, et al., 1981) is reproduced as the first 2 columns of Table 4-1. The table lists the number of responses in time interval classes that end with a failure. The third column records the cumulative number of responses at the end of the time intervals, N_i. Thus, the probability that a response has been made by the class interval time may be estimated by the ratio, N_i / N. A plot of this ratio versus time would produce a discrete, cumulative distribution function (CDF) for the scenario simulated.

Table 4-1
Simulator Response Time Data and Statistics

End of interval t_i (min)	Number of responses n_i	Cumulative responses $N_i = S_i\ n_j$	Number non-responses $N-N_{i-1}$	Discrete distribution $N_i/(N+1)t$
2.2	1	1	10	0.09
3.2	1	2	9	0.18
5.2	1	3	8	0.27
7.0	1	4	7	0.36
8.8	2	6	6	0.55
11.7	2	8	4	0.73
16.7	1	9	2	0.82
31.7	1	10	1	0.91
	N = 10			

Unlike the hardware time-dependent model, the random variable is the elapsed time for a **successful** response, T. In this case, reliability can be expressed as:

$$R(t) = \Pr[T \le t]. \qquad (3)$$

i.e., the probability that a success is achieved by time, t. Since the discrete CDF described above approximates (3),

$$R(t) = F(t), \qquad (4)$$

and its underlying probability density, f, is related to the CDF by:

$$F(t) = \int_0^t f(x)\, dx \,. \tag{5}$$

Fitting the discrete CDF to a continuous CDF amounts to identifying this underlying density, f.

An initial attempt in identifying the distribution of the response time for the data in Table 4-1 is to fit the data using probability paper, i.e., to try to fit the data to a normal (Gaussian) distribution. Since probability paper has no value for 100% response, the ratio mentioned previously cannot be used. Instead, a discrete, cumulative distribution function can be generated using (Bain, 1978):

$$F_d(t_i) = N_i / (N+1) \tag{6}$$

where

N_i is the number of responses by time, t_i
N is the total number of respondents.

The resulting plot is depicted in Figure 4-1. The data does not appear to be Gaussian, as indicated by the curvature of the data when compared to a straight line. However, the downward concavity of the data on this plot suggests using the logarithm of time. This is an alternative fitting scheme also suggested by Swain

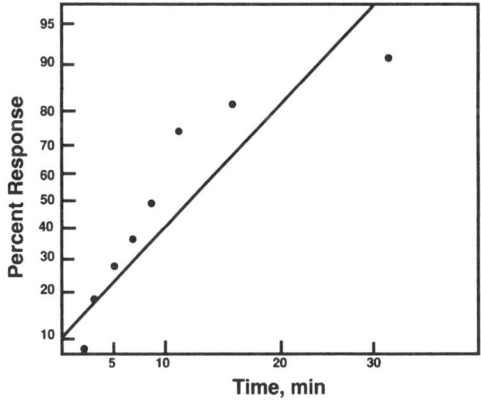

Figure 4-1. Plot of response data on probability paper.

(Swain and Guttmann, 1983) and was used in the ORNL work. Such a fit uses log-probability paper, i.e., the fit of the data is tested against a lognormal distribution. Table 4–2 shows the relationship between lognormal statistical forms and those for the normal distribution. Table 4-3 relates the defining parameters of a lognormal distribution, μ and σ, to more intuitive response time parameters, the median (50%)

Table 4-2
Relating Lognormal to Normal Functional Forms

<u>Standard Normal Form</u>

density $\quad \phi(t) = \exp\left[-t^2/2\right] / [\sqrt{2\pi}] \qquad -\infty < t < \infty$

distribution $\quad \Phi(t) = \int_{-\infty}^{t} \phi(\xi)\, d\xi$

<u>Normal Form</u>

density $\quad f(t,\mu,\sigma) = \exp\left[-(t-\mu)^2/2\sigma^2\right] / [\sigma\sqrt{2\pi}] \qquad -\infty < t < \infty$

distribution $\quad F(t,\mu,\sigma) = \int_{-\infty}^{t} f(\xi,\mu,\sigma)\, d\xi$

Thus, $\quad f(t,\mu,\sigma) = \phi\left[(t-\mu)/\sigma\right]/\sigma$

<u>Lognormal Form</u>

density $\quad f(t,\mu,\sigma) = \exp\left[-(\ln t-\mu)^2/2\sigma^2\right] / [\sigma t\sqrt{2\pi}] \qquad -\infty < t < \infty$

Thus, $\quad f(t,\mu,\sigma) = \phi\left[(\ln t-\mu)/\sigma\right]/\sigma t$

and,

distribution $\quad F(t,\mu,\sigma) = \Phi\left[(\ln t-\mu)/\sigma\right]$

hazard $\quad h(t) = \phi\left[(\ln t-\mu)/\sigma\right] / \left\{\sigma t - \sigma t\, \Phi\left[(\ln t-\mu)/\sigma\right]\right\}$

response time, m, and its error factor, f. Log-probability paper has on one axis the same arrangement of probabilities as does the probability paper while the other axis is the (natural) logarithm of time. The discrete CDF of responses can be generated as in (6) and the points can be least-squares fit as a straight line to a distribution with lognormal density.

The result of fitting the data in Table 4-1 to a lognormal distribution is shown in Figure 4-2. The data seem to fit a straight line somewhat better than for a normal

Table 4-3
Relating Lognormal Parameters to Response Time Parameters

Beginning with the variate, t_p, and using the formula in Table 4-2, yields:

$$p = \int_0^{t_p} f(t, \mu, \sigma)\, dt = \Phi[(\ln t - \mu)/\sigma]$$

or solving for t_p:

$$t_p = \exp[\mu + \sigma \Phi^!(p)]$$

where

$$y = \Phi^!(x) \quad \text{means the unique y such that} \quad x = \Phi(y).$$

Thus, substituting in the median response time, $m = t_{.5}$, yields:

$$\begin{aligned} m &= \exp[\mu + \sigma \Phi^!(0.5)] \\ &\exp[\mu + \sigma(0)] \\ &\exp[\mu] \end{aligned}$$

or $\mu = \ln[m]$.

Similarly, an error factor of f means that the 95th percentile response time, $t_{.95}$, is:

$$\begin{aligned} t_{.95} &= \exp[\mu + \sigma \Phi^!(0.95)] \\ &\exp[\mu + \sigma(1.645)] \\ &\exp\mu \times \exp(1.645\sigma) \\ &m \times \exp(1.645\sigma) \end{aligned}$$

Thus,

$$f = t_{.95}/m = \exp(1.645\sigma)$$

or, $\sigma = \ln(f)/1.645$.

Finally, for the example: $m = 8$ and $f = 3$ implies that $\mu = 2.079$ and $\sigma = 0.668$.

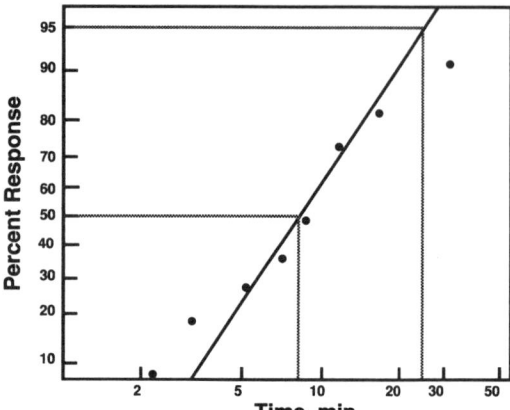

Figure 4-2. Plot of data on log-probability paper.

distribution. Inspection of the curve results in a median time of about 8 minutes and a 95% response time of about 24 min. This results in an error factor of 3 ($= t_{.95}/t_{.5}$ = 24/8) and the lognormal density parameters can be calculated, using Table 4-3, as $\mu = 2.079$ and $\sigma = 0.668$.

Other distribution types have been used for response time correlations—the exponential (Fleming, et al., 1979) and Weibull (Hannaman et al., 1984) distributions. The lognormal family, however, has several nice properties. The first property is that it is a two parameter family, unlike the exponential. The parameters—μ and σ—can easily be replaced using Table 4-3 by the median response time, m, and the error factor, f. These parameters more intuitively specify the central tendency and its dispersion while also being meaningful relative to the notion of response time. One result of requiring two-parameters to define a distribution is that the central tendency can remain constant for differently dispersed data. This could be used as a way to reflect the influences on the uncertainty of a response time. For example, a widely dispersed distribution might mean that the personnel are not highly experienced or that the action requires a decision to be made under conflict or some other source of burden (see Chapter 12).

Another property of the lognormal family is exhibited in its hazard function. A hazard function for the hardware time-dependent model is similar to the failure density. The hazard function is defined as:

$$h(t) = f(t) / (1-F(t)), \tag{7}$$

which can be interpreted as the rate of failures relative to the number of survivors. For the human response time model, f is a rate of success and 1-F is the number of remaining non-(successful) respondents. A discrete density and hazard function can be calculated from data such as in Table 4-1 by:

$$f_d(t) = [n(t_i) / N] / [t_i - t_{i-1}] \qquad t_{i-1} < t \le t_i \qquad (8)$$
$$h_d(t) = [n(t_i) / (N + 1 - N(t_i))] / [t_i - t_{i-1}] \qquad t_{i-1} < t \le t_i \qquad (9)$$

where
  $n(t)$ is the number of successful respondents at time, t
  $N(t)$ is the cumulative number of successful respondents at t, and
  N is the total number of respondents.

Note that for any i:

$$\sum \{f_d(t_i) \times (t_i - t_{i-1})\} = N_i / N \qquad (10)$$

which was the initial intuitive estimate for the discrete CDF. This means that the relationship between f and F in (5) will be approximated when using (8). Also, since the empirically derived, discrete density function, $f_d(t)$, consists of step functions, i.e., is a histogram, its integral, the CDF, is a piecewise-continuous function consisting of ramp functions, i.e., the curve is linear from the point at t_{i-1} to the point at t_i.

Since equation (9) is the ratio of the number of responses to those left to respond over a time interval, this hazard function in the human reliability case is, thus, really a solution rate.

The difference in (8) and (9) is the normalizing factor, N versus n. As a result, the density function can be thought of as indicating the *overall speed* of the responses, whereas the solution rate (the hazard function) indicates the *instantaneous speed* of the solutions (as in Shooman, 1987). Table 4-4 shows the discrete, density, its distribution function, and the non-response or hazard function for the data in Table 4-1. These values are compared to those for a lognormal density, distribution, and hazard function with median of 8 minutes and an error factor of 3. Figures 4-3 through 4-5 show graphically the density, distribution, and solution functions, both empirically produced by the equations and data above and calculated for a lognormal distribution with median of 8 and error factor of 3.

The solution rate is suggestive of a real phenomenon. This curve, for lognormal families with σ greater than about 0.4, has a maximum solution rate. This behavior has also been noted in repair troubleshooting responses (Wohl, 1982) where an interpretation was suggested for this phenomenon. The operator(s) first try standard heuristics, or rules-of-thumb, that have worked in the past. This pattern matching

Table 4-4

Discrete and Calculated Non-Response Statistics

Data		Discrete			Calculated		
t_i	N_i	f_d	F_d	h_d	f	F	h
2.2	1	0.05	0.10	0.05	0.04	0.03	0.04
3.2	2	0.10	0.20	0.11	0.07	0.08	0.08
5.2	3	0.05	0.30	0.06	0.09	0.26	0.12
7.0	4	0.06	0.40	0.08	0.08	0.43	0.15
8.8	5	0.11	0.60	0.09	0.07	0.56	0.16
11.7	7	0.07	0.80	0.17	0.04	0.72	0.14
16.7	9	0.02	0.90	0.20	0.02	0.86	0.14
31.7	10	0.01	1.00	0.07	0.00	0.98	0.11

will catch most situations fairly quickly. However, some situations will be out of the experience base of some crews and the problem at hand must be solved by more knowledge- or first-principle-based mechanisms. These mechanisms are slower, which is the source of the decrease in reliability*. Below a σ level of about 0.4 (or an error factor of 2), the convexity property seems to disappear. This could be interpreted to mean that the situation is well-within the crew's experience, which is the source of the smaller dispersion as indicated by σ.

Finally, since F(t) is the probability that a successful response will occur at least by time, t, 1-F(t) is the non-success or failure probability at t. This complementary cumulative distribution function (CCDF) is used most in HRA as a time reliability correlation (see Chapter 10). Figure 4-6 shows the TRC that results from the data.

In summary, time-dependent human failure reliability mathematics has a structure analogous to that for hardware. However, the concept for HRA is quite distinct at the implementation detail level, as Table 4-5 indicates.

Time-Independent Human Failure Models

There has long been a controversy in hardware reliability analysis as to whether there are truly any time-independent, or demand failures. The argument against such failures is that a demand merely acts as a discovery mechanism for a failure that occurs over a standby duration. It turns out that this is not a convincing argument from either a data analysis or a mechanism point-of-view, but this is not relevant

---

* These mechanisms are less reliable in the sense of (3). However, these cognitive mechanisms are the **only** means by which people can solve **some** problems.

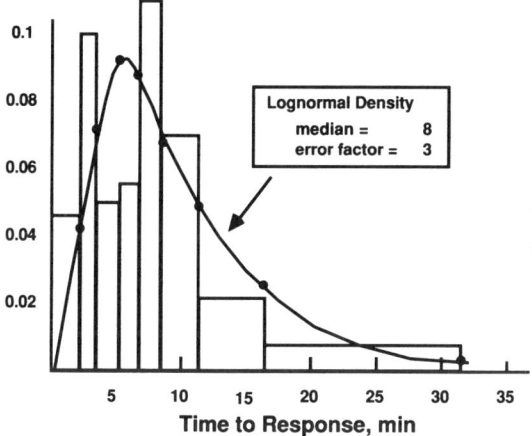

Figure 4-3. Density for the data of Table 4-1.

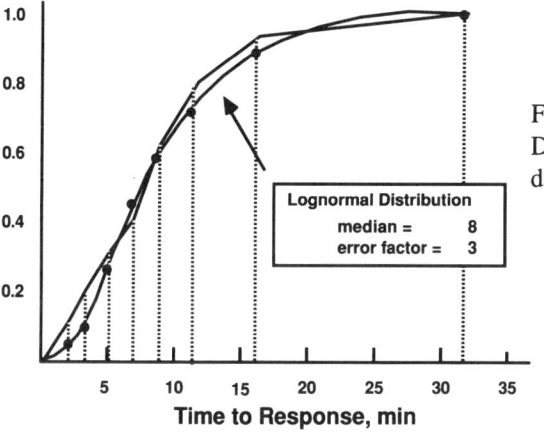

Figure 4-4. Distribution function of data in Table 4-1.

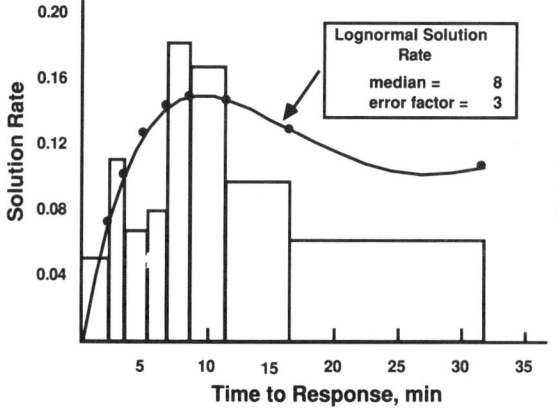

Figure 4-5. Solution rate of data in Table 4-1.

here. In the case of human reliability, there are compelling reasons for time-independent failures.

Table 4-5
Comparing Time-Dependent Reliability for Human and Hardware Failures

	Hardware	Human
f(t)	failure density	success density
F(t)	$\int_{-\infty}^{t} f(x)\,dx$	$\int_{0}^{t} f(x)\,dx$
R(t)	$\Pr[t \geq t] = 1 - F(t)$	$\Pr[T \leq t] = F(t)$
$\lim_{t \to \infty} R(t)$	0	1

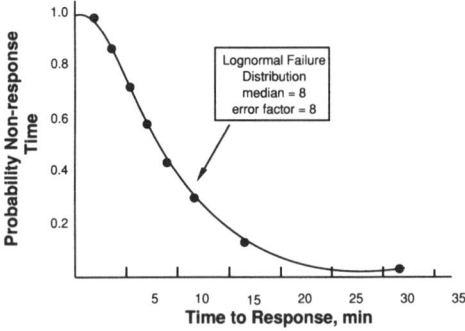

Figure 4-6. The time reliability correlation from the data in Table 4-1.

Many, if not most, fundamental human processes are fast—on the order of tens to hundreds of milliseconds (Card, et al., 1983, pp. 25-86). As a result, failure mechanisms cannot accrue much time to fail in the sense of (3). Also, the study of errors that are actions-not-intended (Reason and Mycielska, 1983) shows non-temporal processes at work.

Generally, a time-independent failure occurs following the establishing of a goal and the making of a plan to achieve the goal, i.e., the goal is the demand. At this point, habit or procedure usually takes over to implement the plan. Such slips and lapses manifest themselves in one of three main ways (Reason and Mycielska, 1983, pg. 25):

1. omission—a step in the plan is not performed
2. mis-selection—in performing a step, the wrong object is used or manipulated
3. repetition—a step is performed again unnecessarily.

Omissions and repetitions are typically "place-losing" errors (pg. 138-140) due to a lapse in memory or faulty attention. Mis-selection seems to occur as the result of the so-called task capture phenomenon mentioned earlier. Either the person is distracted by unanticipated external events or preoccupied by emotional states or unrelated but interesting ideas, and reverts to more frequently or recently practiced similar tasks (pg. 60-61).

The model for reliability in time-independent, i.e., plan-directed, behavior is:

$$R = Pr[s \leq \text{ capacity of the object}] \tag{11}$$

i.e., the probability that some slip causal agent does not overwhelm the attentional resources allocated. This probability is estimated as the number of failures of a given type divided by the number of demands that could have led to that type of failure.

The analogy between human and hardware reliability begun in Chapter 3 and continued in the previous section can be further drawn as in Table 4-6. Standby failures in human performance come from failures in vigilant behavior, i.e., times when the operator is doing nothing but baby-sitting the process control board or monitoring a single or several indicators. This latter behavior has little if any risk-significance in nuclear power plants because of the constant workload from other sources, such as regulatory or maintenance activities. Demand failures in human performance occur when a plan is known and well-established and the implementation of the plan is required. Failures in plan-directed activities will be seen in normal and off-normal activities. Finally, "running failures" in human performance come when the events ongoing are so uncertain or unanticipated as to require significant diagnosis, decision making, and/or on-the-spot planning, which are slow, usually conscious, processes that have time-dependent characteristics.

The next chapter discusses the data that is available to support the reliability formulation of this chapter.

Table 4-6
Continuing the Human and Hardware Analogy

Hardware failure	is analogous to	Human activity
standby failure demand failure running failure		vigilant behavior plan-directed (task) behavior event-driven behavior

Chapter 5
HRA DATA

The trail from "raw" data to useful information is laborious and liberally sprinkled with human judgment (see Figure 5-1). So natural is the capacity for interpretation, that people often forget how much they impose on what reality provides. HRA data is not only woefully sparse but is not usually presented in its raw form, leaving questions for its users as to its pedigree and utility. The next section addresses the current sources of HRA-like data and makes some judgment as to their pedigree. The latter sections discuss the data that is packaged in the technique for human error rate prediction (THERP, see Chapter 6) and the data that is becoming available from nuclear plant simulators.

Figure 5-1 shows a trail from raw data to information. In reliability engineering, raw data takes the form of failure records, opportunity counts, and attribute lists. Users of raw data typically select among the data according to some set of attributes and calculate the failure statistics for that "population". The result is compiled data. Further judgment as to outliers and the discriminating attributes are provided a full-fledged data analysis to fit the data to distributions or determine statistical parameters. The result can then be called interpreted data. Massaging this interpreted data with theory finally leads to something that is informative and can be acted upon.

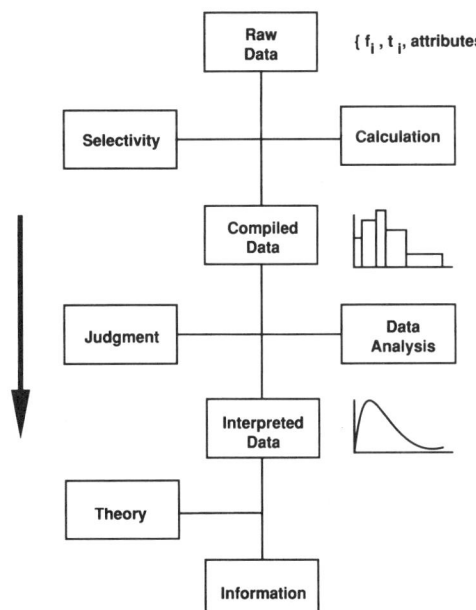

Figure 5-1. The trail from raw data to information.

39

While there are hardware reliability analysts who are data "starved", these analystst are usually involved in developmental or high-technology industries (Fragola, 1983). In these cases, the volatility of the data sets developed (which means how soon after the data become available do they become obsolete) causes the user to continually seek out new information. However, in many cases, hardware reliability analysis has the luxury of having a relatively strong basis in data. When data is plentiful, then statistical techniques can be applied as can empirical models to supplement areas where the theory of failure mechanism or causation is lacking. In cases where the occurrence of a failure is a rare event, (i.e., the data that are available are too few to make estimates of the events' occurrence probabilities) and its frequency is thereby not well-established historically, then estimates of the frequency of occurrence can be made by using data from more frequently occurring (often less severe) events. Also, aggregation of data can be made on devices which can be considered to some degree similar, provided proper attention is paid to the tolerance of the data.

The problem with using such a data analysis approach in HRA quantification is that, for HRA, aggregation of data is often inappropriate. Human failure events differ so much in the circumstances and mechanisms leading to failure, that each one seems unique, and therefore confining events into a statistical population can produce tolerance problems which are often unacceptable. In fact, in some cases, the coverage becomes so broad, the the data loses all applicability (Brune, et al., 1983). Despite these problems, HRA data has been collected since the 1960s. Collection has been greatly increased over the last several years so that now patterns in the data are just being realized. The development of data patterns and the consequent development of aggregated data with reasonable tolerance bounds is one of the primary goals of HRA technology in the 1980s—as it was for hardware reliability technology in the 1960s.

Development of Existing Data Sets

The data needs from HRA have only recently become more focused as the methodologies have become more structured and standardized. Originally, HRA technology was seen as an outgrowth of human factors technology and therefore the data sets developed were structured on a task unit basis, as in done in a procedural task analysis. The model underlying this tactic considered the human as an implementor of pre-programmed activities, i.e., a "button pusher", a "knob turner", or a "procedure implementor". This model of human interaction led to the creation of "data cells" which included estimates of the probability that the operator could manipulate a given piece of equipment on demand, or implement a given procedure correctly. Earlier, interest in the man/machine interface was mostly confined to the military (Munger, et al., 1962) and the early data sets reflected this perspective. As the nuclear power industry sought to introduce quantitative estimates of human performance into their safety analyses (e.g., USNRC, 1975), expert judgment was used to extrapolate the existing body of military experience to commercial nuclear plant operations and resulted in the publication of a handbook (Swain and Guttmann, 1983) which codified this extrapolation.

Further investigation into human performance in the commercial nuclear power plant setting, especially after the TMI, Unit 2 incident in March 1979, caused a change in the perception of the data needs. The model underlying this changed perception considered the human more as a system analyzer and decision-maker. Simulator training sessions began to illuminate more clearly the role the operator played as an active participant in accident sequence evolution and resolution. These sessions also indicated the potential for gathering higher-level decision making data from the simulator. Additionally, continued implementation of the more procedurally-oriented approaches led to a desire for a more structured data interface to improve the consistency in data application (Fragola, 1983). This work also pointed out that human interaction with equipment, procedures, and control panels could be condensed into broader categories with reasonable uncertainty bounds—thereby eliminating the need from numerous detailed data cells.

As a result of the meetings at Myrtle Beach in 1979 (Schmall, 1980) and in 1981 (*Conference Record*, 1982), two distinct classes of human interactions began to emerge—one focusing on the original view of man as a manipulator—and the other concentrating on man as an information processor and decision maker in a time-constrained environment. These two considerations implied very different data needs. Whereas, the expert judgment, time-independent data developed early on could still be considered applicable to the determination of the probability of equipment manipulation, a totally new time-dependent data set needed to be developed to estimate the probabilities of human decision making during accident sequences.

1982 Brookhaven National Laboratory Study - NUREG/CR-3010

Soon after the 1981 Myrtle Beach Conference had clearly indicated the need for time-dependent HRA data, a series of simulator exercises were conducted by Oak Ridge National Laboratory (Bott, et al., 1981 and Beare, et al., 1982). The former of these studies investigated PWR sequences and the latter, BWR sequences. The early study presented cumulative operations team response time data in a log-probability format. This presentation indicated that in a majority of the accident sequences investigated, the data could be reasonably well fitted to a straight line in this presentation format (see Figure 4-2). Concurrently, a research program was undertaken at Brookhaven National Laboratory. One the goals of this effort was to investigate potential mathematical models for characterizing time-reliability correlations. The research indicated the lognormal distribution as a potential candidate because it satisfied the requirements of a response time density function and the corresponding cumulative distribution function could easily be characterized and presented in a linear fashion on log-probability paper. The final report (Hall, Fragola, and Wreathall, 1982) also presented all the results of the PWR studies on a single figure. This figure presented the first evidence of the existence of various response time probability "regimes" and the potential characteristics of the distributions which lie in each of these regimes. This early collection of data has been shown to be consistent with later results and has provided primary reference information for later human reliability research.

1984 LER Study

The realization of the importance of human error led to an attempt to use existing data sets as a resource to establish human error rates. To this end, a study was conducted in 1984 (Voska and O'Brien, 1984) to investigate the set of plant Licensee Event Reports (LERs) to determine their potential usefulness as a source of human error data. Practicality, acceptability, and usefulness were considered as criteria. The study determined that only 2% of all LERs contain human error information, which could be useful in the estimation of human error probabilities (HEPs). Also, a survey of experts indicated that LERs could be somewhat useful as a data source, but that this usefulness was limited by the nature of LER reporting. Finally, the report concluded that, since LER data was most applicable to components, it would be of limited usefulness to determine human participation at the accident sequence level, where most risk-relevant events seem to take place.

1984 HRA Data Survey

In the 1984-85 time frame, studies were conducted to evaluate the applicability of the data sets available at that time to the newly-defined HRA needs. One project (Fragola, 1984) attempted to review all the sources available that addressed human error or human reliability. The sources identified were catagorized by their relevance to either procedural or decision-making data needs, and were then reviewed for their applicability to the requirements of quantification. In the case of data sets relevant to procedural tasks, the data sources were classified by data origin (judgment, simulator, in-plant), whether there was a clear indication of data pedigree and applicability in the data cell definition, and by whether the actual man/machine interface was considered (i.e., selection of a pump over a valve involves confusion in the selection between switches on a control panel). When decision-making tasks were addressed, the evaluation considered whether a time-reliability correlation approach was used, what the data origin was, and whether crew size and experience level were addressed.

The study results indicated that, of the 49 documents reviewed, only two sources met the criteria for procedural task qualification (one based upon judgment, and the other on plant data), and only one source provided data for decision-making task qualification (from simulator results).

1985 Nordic Project Work

During the time period of 1981-1985 the Nordic Liaison Committee for Atomic Energy investigated the influence of human error in test and calibration activities conducted in nuclear power plants. This work reviewed problems related to the optimization of test intervals, organization of test and maintenance activities, and the analysis of human error contribution to the overall risk in test and maintenance tasks. While the published report (Anderson, and Liwång, 1985) contained little HRA data (recorded data was limited to one table which presented a qualitative classification of 54 errors observed in testing and maintenance), the report did indicate the need for a "consistent theoretical framework" to "increase the explantory power of data from studies of Human Error in maintenance tasks".

1985 Brookhaven National Laboratory PRA Review

This study included a review of some 19 PRA studies of U.S. commercial nuclear power plants which had been completed and published as of late 1984. The study attempted to identify, collect, and store all the HRA data included in these studies. For each HRA/PRA datum identified, all descriptive information presented in the PRA concerning that datum was entered into the data record. This descriptive information included data concerning personnel, actions taken, performance shaping factors (PSFs) used, situation description, and indication of the systems involved. Table 5-1 which was reproduced from the final report (O'Brien and Spettell, 1985) indicates the PRAs included in the analysis.

The study found 1,976 HRA/PRA data records which were recorded. Of the records recorded, 78% concerned operator actions, roughly 10% concerned maintenance personnel actions, 10% instrumentation and control personnel actions,

Table 5-1
PRAs Included in BNL HRA Project
(From O'Brien and Spettell, 1985)

Name and Type of Plant	Sponsor	PRA Level*	Contractor
Arkansas Nuclear One Unit 1, PWR 2-loop	NRC (IREP-II)	2	SNL, BCL, SAIC
Calvert Cliffs Unit 2,	NRC (RSSMAP)	2	BCL, SNL, EA
Calvert Cliffs Unit 2,	NRC (IREP-II)	1	SAIC
Crystal River Unit 3,	NRC (IREP-1)	2	SAIC
Indian Point Units 2 & 3	Power Authority of NY	3	Consolidated Edison of NY and Power Authority of NY
Midland Power Plant Units 1 and 2, PWR	Consumers Power Company	3	PL&G
Oconee Unit 3, PWR 2-Loop	NRC (RSSMAP)	2	SNL
Seabrook Station Unit 1	Public Service Co., Yankee Atomic Elec. Co.	3	PL&G
Sequoyah Unit 1, PWR 4-Loop	NRC (RSSMAP)	2	SNL, BCL
Surry Unit 1, PWR 3-Loop	AEC/NRC (WASH-1400)	3	N. Rasmussen, MIT

* PRA level refers to the extensiveness of the PRA methodology. Level 1 PRAs include an analysis of events and systems in relation to core-melt processes. Level 2 PRAs include an analysis of radionuclide release and transport as well as an analysis of core-melt processes. Level 3 PRAs include an analysis of core-melt processes, an analysis of radionuclide release transportation, and an analysis of environmental transport and consequences.

Table 5-1, Cont.

Yankee Rowe, PWR Electric Co.	Yankee Atomic	-	EI, Yankee Atomic Electric Co.
Zion Units 1 and 2, PWR/4-Loop	Commonwealth Edison	3	PLG, Westinghouse
Big Rock Point (BWR) Co.	Consumers Power	3	Consumers Power Co.
Browns Ferry Unit 1, BWR/4	NRC (IREP-II)	2	EG&G Idaho, EI
Grand Gulf Unit 1, BWR/6	NRC (RSSMAP)	2	SNL, BCL
Limerick Generating Station Units 1 and 2, BWR/4	Philadelphia Electric Co.	3	Philadelphia Electric Co., GE Co., SAIC
Millstone Unit 1, BWR/3	NRC (IREP-II)	2	SAIC
Peachbottom Unit 2, BWR/4	AEC/NRC	3	N. Rasmussen, MIT
Shoreham Nuclear Power Station, Unit 1, BWR	Long Island Lighting Co.	3	SAIC

Acronyms

BCL - Battelle Columbus Laboratory
EA - Evaluation Associates
EI - Energy Incorporated, Inc.
IREP - Integrated Reliability Evaluation Program
MIT - Massachusetts Institute of Technology
NRC - Nuclear Regulatory Commission
PL&G - Pickard, Lowe, and Garrick
RSSMAP - Reactor Safety Study Methodology Application Program
SAIC - Science Applications International Corporation
SNL - Sandia National Laboratories

and the remainder were shift supervisor and shift technical advisor-related. Specific actions were identified for all the records. It was found that 48% concern operating actions, 14% concern testing, 12% concern maintenance, 9% concern calibration and the remainder concern other actions. Only 59% explicitly considered PSFs. Of these, 32% were related to the use of procedures, 26% related to stress, 14% to time available, and the remainder to other PSFs.

Although this study includes a substantial amount of information it should be remembered that the PRAs included are all pre-1985 and thereby primarily used 1982 technology in their approach to HRA. At this point in time HRA approaches were still developing rapidly and so it should not be surprising that the study found that:

> "Very little evidence of consistent consideration of specific systems or accident sequences was observed. Instead, different PRAs modeled very different systems in detail. Similarly, aside from LOCA-type accidents, no consistent set of sequences appeared to be analyzed."

The study also found where it compared the data available to the data required to address Generic Safety Issues (judged to be of import to NRC), that only 17% of the HRA needs could be addressed by the then current PRA/HRA data. As the study concluded: "Overall, the currently available HRA/PRA data address less than 17% of all data needs arising from the list of working level issues."

1986 Peach Bottom ATWS Sequence HRA

This study provided the Brookhaven National Laboratory input to the Peach Bottom Accident Sequence Evaluation Program (ASEP) analysis of anticipated transient without scram (ATWS) sequences with main steam isolation valve (MSIV) closure. BNL conducted a task analysis to develop an event tree for this event and quantified the branch points using what were considered to be appropriate HRA methods. Human error probability data and uncertainty bounds are given in the report (Luckas et al, 1986) along with a comprehensive discussion of how they were developed.

The data developed were derived from the previous BNL studies—time-dependent data (O'Brien and Spettell, 1985) and time-independent data (Hall et al, 1982). The derived data were selected by consensus of BNL analysts and were modified according to PSFs, after taking into account the outcome of previous events in the sequence, using a structured expert judgement technique (Embrey, et al., 1984). The PSFs considered included both human traits and conditions of the work setting likely to influence an individual's performance.

1987 Idaho National Engineering Laboratory NUCLARR Library

The Nuclear Computerized Library for Assessing Reactor Reliability (NUCLARR) is an automated data base management system used to process, store, and retrieve, human and equipment reliability data in a ready-to-use format. NUCLARR was developed to provide the risk analysis community with a repository of human error and harware failure rate data that can be used to support a variety of analytical techniques to assess risk. The human error component of the NUCLARR System complies with the specifications and procedures documented in an earlier study (Comer and Donovan, 1985).

The human error data library access system described in the final report (Gertman et al, 1987) may become a powerful aid to human reliability analysts. However, at present, the only data sets it includes are the THERP data sets (Swain and Guttman, 1983) for time-independent data, and the Oconee PRA (NSAC, 1984) for time-related data. Although other data sets are to be included, the data library will continue to be over-burdened with the extensively detailed human error data cells that were developed earlier (Topmiller, et al., 1982, Comer, et al., 1983, and Comer, et al., 1985). These numerous data cells are no longer considered useful by current PRA practitioners. This unnecessarily detailed approach, and its associated multitude of empty data cells, is consistent with the early approaches taken toward the analysis of human reliability. However, it fails to support the current emphasis, which tends to concentrate on cognitive failures.

Data in THERP

THERP (see Chapter 6) analyzes human actions into discrete task steps which might be practiced in similar ways in different task environments. As a result, there was some hope that data could be found for the component probabilities from other industries and compiled into an HRA data set. The data at any one facility would necessarily be rare as the following example demonstrates.

Suppose a bomb assembly task (the environment that motivated the development of THERP) consists of ten steps, each of which is checked by another worker. Then, the task failure probability reduces to summation of ten step probabilities, each of which is the product of a basic failure probability and the dependent failure probability of the failure of the checker:

$$\Pr[\text{task failure}] = 10 \times P_{basic} \times P_{dependancy}. \qquad (1)$$

Let's assume that the task is performed daily, five days a week, fifty weeks a year and that there has been no task failure in ten years of facility operation. The probability of seeing zero step failures in 2500 opportunities is 1.33×10^{-4}, or one in 7500. This means that

$$P_{basic} \approx \sqrt{(1.33 \times 10^{-5})}, \quad \text{or} \quad 3.7 \times 10^{-3.} \tag{2}$$

Presuming the dependency step probability to be ten times as large as the basic probability, a new estimate is:

$$P_{basic} \approx 1.2 \times 10^{-3}$$
$$P_{dependency} \approx 1.2 \times 10^{-2}. \tag{3}$$

This, in fact, is a typical value for a basic probability in the THERP data set as Table 5-2 indicates.

As noted before, the data used in THERP does not come from nuclear power plant tasks but is "derived" from data from other industries (for example (Munger, et al., 1962)). Nuclear plant simulator data is slowly accruing and may someday supercede the THERP data (see Whitehead, et al., 1987).

Each basic failure probability can be adjusted in THERP by so-called performance shaping factors. Data relevant to estimating these influences on task actions is even more sparse, although THERP provides some guidance relative to the needed a values. Techniques incorporating direct subjective probability estimation (Comer, et al., 1984) or indirect use of expert judgment (Embrey, 1984 and Phillips, et al., 1984) have been used as surrogates to these data needs.

THERP also provides a time-dependent model of the diagnosis process, called the nominal diagnosis model. This time-reliability correlation (TRC) is supposed to apply to estimation of the failure probability of choosing the proper procedure, given a plant upset. This TRC was derived from the so-called NREP procedures guide (Papazoglou, et al., 1984) and is a consensus of the judgment of several risk and HRA analysts, and again is not the result of data. The nominal diagnosis model TRC of THERP is discussed in Chapter 10.

Data from Nuclear Plant Simulators

A growing source of data useful to HRA are the simulators owned by some utilities and government facilities. Simulator data is the basis of several HRA methods for quantifying time-dependent human failures. The time reliability correlations to be presented in Chapter 10 were based in part on data collected in an NRC-sponsored

Table 5-2
Some Typical Values from THERP
(From Swain and Guttmann, 1983)

Failure Event	HEP*	EF*	Table
omitting step in procedure	0.003	5	20-5
fail to use test or calibration procedure	0.05	5	20-6
omission in procedure, with checkoff, ≤10 items	0.001	3	20-6
omission in procedure, with checkoff, >10 items	0.003	3	20-6
omission in procedure, without checkoff, ≤10 items	0.003	3	20-6
omission in procedure, without checkoff, >10 items	0.01	3	20-6
commission in reading digital meter	0.001	3	20-10
commission in reading analog meter	0.003	3	20-10
commission in reading chart recorder	0.006	3	20-10
inadvertent operation of manual control	plant-specific		
select wrong control, controls labelled only	0.003	3	20-12
select wrong control, controls in functional grouping	0.001	3	20-12
select wrong control, mimic	0.0005	10	20-12
turn 2-position control wrong way	0.0005	10	20-12
" , population steorotype violated	0.05	5	20-12
select wrong circuit breaker, densely packed, labels	0.005	3	20-12
select wrong local valve, similar items, clear labels	0.001	3	20-13
select wrong local valve, similar items, unclear labels	0.005	3	20-13
checker fails to find error, routine, with procedures	0.1	5	20-22
checker fails to find error, routine, special activity	0.01	5	20-22
checker fails to find error, routine, safety import	0.001	5	20-22

* HEP stands for human error probability and EF is the lognormal error factor on the assumed distribution on the HEP.

program on allowable operator times-to-response as discussed previously. This program was conducted by the Oak Ridge National Laboratory (ORNL) in the late 1970s and early 1980s, using the Sequoyah and Brown's Ferry simulators in the Tennessee Valley Authority's simulator facility (Bott, et al., 1981 and Beare, et al., 1982). Lately, data is beginning to become publicly available from the work performed by the Electricité de France (Villemeur, et al., 1987) and by a joint program of the Electric Power Research Institute and the NRC. The US work used the simulator at the LaSalle nuclear power plant (Whitehead, et al., 1987).

A simulator typically can model some set of postulated accident sequences beginning with a type of with a type of plant upset (initiator), then modeling one or more system casualties for a specified time into core cooling degradation, until the simulation fidelity becomes too uncertain to continue. The newer simulators are quite flexible and can model sequence phenomena including core melt. The older models are not so flexible nor accurate.

An unresolved question is how accurate a simulated exercise is relative to what might happen in an actual situation of a similar type. There is no apriori answer to this question, e.g., a simulator exercise could elicit more reliable behavior because the crew anticipates the coming test or the behavior could be less reliable because they know it is just a test. The ORNL work attempted to correlate field data to the simulator tests but the results did not conclusively indicate a uniform "bias".

A simulator exercise typically consists of the following steps:

1. Identify the sequence to be modeled, including the initiator and the equipment status. This sequence is usually chosen to be representative of a risk-significant situation or to be important from some other HRA perspective.

2. Choose one or more crews (preferably actual crews, i.e., groups of operators that regularly work together) to perform the exercises.

3. Run the exercise, recording the times when certain pre-chosen types of events, including actions by the crew, occur.

4. Conduct post-exercise interviews to fill in the informational gaps, e.g., the decisions made, the indicators used, etc.

5. Analyze the data.

Data is taken as "time lines" of key events, i.e., and event, time, and comment matrix is constructed. This information is then arranged into time reliability correlation format by focusing on a particular action, determining from the timelines the time from a leading cue to perform the action until the action was performed (or the sequence was stopped), compiling statistics for the action using some aggregation formulation, such as (1), and plotting the resulting data on log-probability paper or some other useful graph paper. Aggregation was performed in the ORNL work according to (Bain, 1978):

$$p_i = N_i / (N+1) \qquad (4)$$

where   i indicates the ith failure of a specified type
and N_i is the cumulative number up to and including the ith
and p_i is the empirical cumulative probability for the time of the ith failure.

The data so derived can also be fitted to any of the typical probability distribution functions or can be plotted on various probability papers. TRCs have been derived using Weibull, log-linear, lognormal, and exponential distribution families. The ORNL, EdF, and LaSalle data (which was compiled by Sandia National Laboratories for the NRC) were all plotted on log-probability paper, i.e., fitted to a lognormal family of distributions. These fits seem to be good and have the characteristics described in Chapter 4.

Figure 5-2 shows some of the ORNL plots, identifying the initiator modeled in the exercise and the resulting lognormal parameters, the median response time and error factor. The ORNL simulators were fairly old and inflexible and, thus, the sequences modeled were not as severe as PRA typically studies. For example, the LOCA sequence was so small a pipe diameter equivalent size that it is often ignored in PRA and the steam generator tube leak was a leak not a rupture as assumed in PRA. This increases the uncertainty in generalizing from the operator responses in these limited scenarios to more PRA-oriented scenarios.

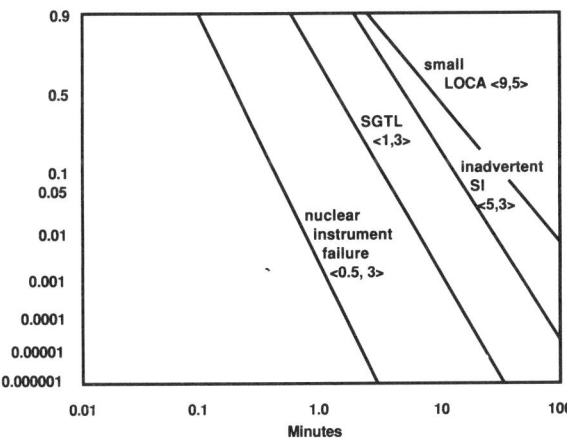

Figure 5-2. Some of the ORNL simulator data. (Interpolated from Bott, et al., 1981.)

Figure 5-3 shows a single curve from the EdF studies in their PWRs. This curve is an aggregation over several initiators—loss of offsite power and one diesel generator; total loss of the residual heat removal system (RHRS); a steam generator tube rupture; loss of the primary coolant when the RHRS is cooling the reactor; total

52 Human Reliability Analysis

Figure 5-3. An aggregation of the EdF simulator data.

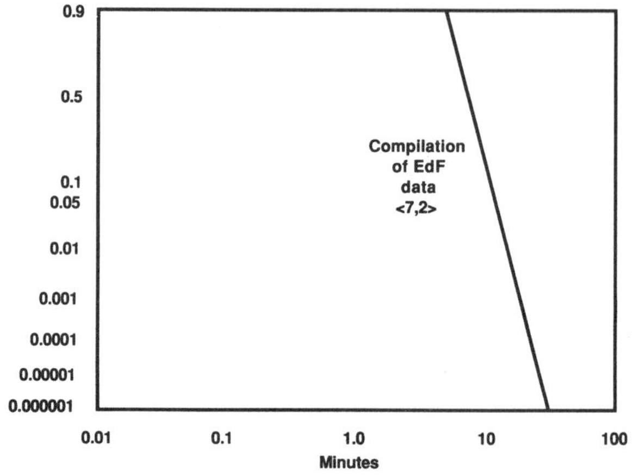

loss of heat sink; and total loss of feedwater to the steam generators. The data are also aggregated over the two series of exercises reported.

Finally, Figure 5-4 shows some non-nuclear data in the short response time regime (about a one sec median response time). This data was taken in vigilance/response tasks (Greene, 1969) in which a person is watching an indicator and attempts to push a button as fast as the indicator lights. It is interesting to note that the more reliable (faster) human response occurs when the frequency of signals is greater, i.e., (up to a limit) more stimulus increases the frequency of response, a phenomenon noticeable in computer arcade games as well.

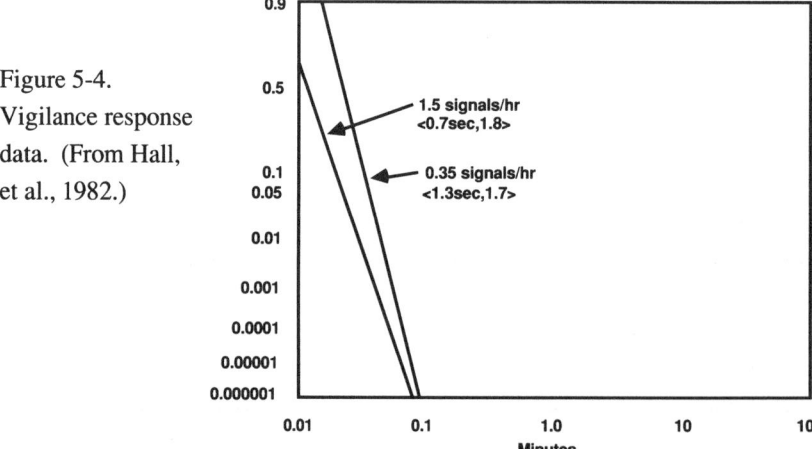

Figure 5-4. Vigilance response data. (From Hall, et al., 1982.)

HRA Data   53

Sandia (LaSalle) Simulator Data

Eight LaSalle simulator drills, or sequences, provided the data for the TRCs developed by Sandia (Whitehead, 1987). The drills consisted of the following sequences:

1. Main steam isolation valve (MSIV) closure followed by failure to automatically scram (ATWS), motor driven feedwater pump available.

2. MSIV closure followed by ATWS, motor driven feedwater pump unavailable.

3. Spurious turbine trip with failure of narrow range level instrument and high pressure injection.

4. Spurious turbine trip with failure of narrow range level instrument, wide range level instrument, and high pressure injection.

5. Loss of offsite power, failure of all diesel generators (station blackout), and failure of the reactor core isolation cooling system (RCIC) injection valve.

6. Loss of offsite power, failure of two diesel generators. The third diesel generator (DG) starts and loads and isolates RCIC. The third DG fails 20 mins into the sequence (delayed station blackout).

7. Transient with dc bus 1A failure, shorting to ground, leading to subsequent failures that threaten critical parameters.

8. Feedwater line break in the steam tunnel, resulting in loss of flow from the feedwater system. Subsequent failures result in loss of all high pressure systems and low pressure systems must be used.

Each of the drills included at least one opportunity for operator action. Sandia then grouped these action events according to similarities in operational characteristics. The result was eleven generic action types. The response time data for each specific action in a group was then aggregated to generate a single TRC for the generic action type. Table 5-3 lists each generic action type, the drills that contribute specific actions to the type and the specific actions.

Figure 5-5 shows the TRCs that result from aggregating the LaSalle data into the eleven general actions types. Table 5-4 identifies the median and mean response times and error factors for each general action type. Note that most of the data falls between two TRCs: one with median response time of 1 min and an error factor of 3 and the other with a median of 20 min and an error factor of 5 (shown as the grey area in Figure 5-5).

The sequences and actions studied by Sandia are not always significant to the risk of a BWR. Typically, however, sequences (drills) 1 and 2, the ATWS

Table 5-3
**Recovery Actions Aggregated into Generic Action Types
LaSalle Simulator Exercises**

1. Operators fail to anticipate automatic actuation of a system or fail to initiate manually actuated system.	Drill 1	initiate residual heat removal after reactor (Rx) trip automatic actuation.
	Drill 2 & 2b	initiate suppression pool (SP) cooling after reactor trip
	Drill 3	initiate RCIC after station blackout
	Drill 4	initiate SP cooling after DG1A loads
	Drill 6	close main steam isolation valves (MSIV) after level 7 alarm
	Drill 8	initiate SP cooling after Rx trip.
2. Operators fail to use low pressure systems with high pressure systems unavailable.	Drill 8	depressurize after RCIC failure
	Drill 8	inject low pressure water after RCIC failure
3. Operators fail to manually operate a component or system which failed to automatically actuate or operate.	Drill 3	send operator to open FO13 after failure of FO13
	Drill 4	reset RCIC isolation after DG1A loads
	Drill 8	request RCIC investigation after RCIC failure
4. Operators fail to restore safety-related inhouse electrical buses or supply equipment.	Drill 3	request DG repair after station blackout
	Drill 4	request DG1B repair after station auxiliary transformer (SAT) failure
	Drill 6	request DG1A investigation after DG1A failure
5. Operators fail to restore offsite-supplied electrical buses or equipment.	Drill 3	request crosstie after station blackout
	Drill 3	request SAT repair after station blackout
	Drill 4	request SAT repair after SAT failure
	Drill 4	request crosstie after SAT failure
	Drill 6	restore bus 151 locally after Rx trip

Table 5-3, Cont.

6.	Operators fail to manually scram the reactor.	All All	mode switch after Rx trip manual scram after Rx trip
7.	Operators fail to manually override a system that automatically functions when the function of the system would challenge a critical parameter.	Drill 1 Drill 4 Drill 6 Drill 8	jumper vent & purge system (VP) after drywell isolation restore VP after drywell isolation restore VP after DCA failure restore VP after drywell isolation
8.	Operators fail to inject standby liquid control system (SLCS) water into the reactor vessel.	Drill 1	inject SLCS water after SP temperature high-high alarm
9.	Operators fail to request for last resort "dirty" water to prevent core damage.	Drill 4 Drill 4	depressurization after station blackout request diesel fire pump after station blackout
10.	Operators fail to locally operate a manually controlled component normally operated from the control room when it fails to operate from the control room.	Drill 2 & 2b Drill 6	send operator to close steam dump valve after scram request attempt request air restoration after service air pressure low alarm
11.	Operators fail to manually override a false control signal when no direct indication exists that the signal is incorrect.	Drill 4	request bypass of RCIC isolation after RCIC isolation because of room overheating

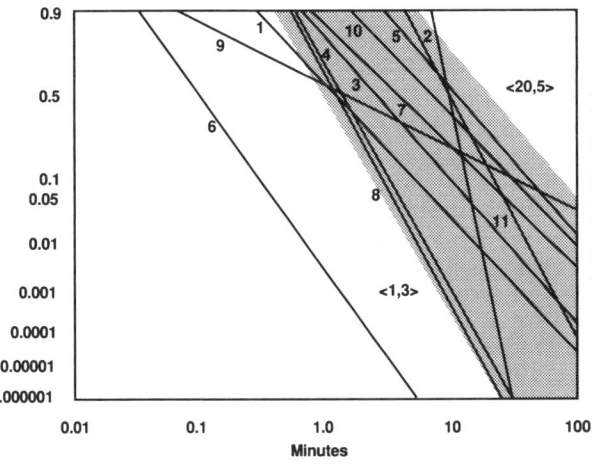

Figure 5-5. TRCs for generic action types from the data from the LaSalle simulator. (Interpolated from Whitehead, et al.,

Table 5-4
TRC Parameters for LaSalle Generic Actions

Drill	Description	Response Time Median (min)	Response Time Mean (min)	Error Factor
1	failure to manually operate prior to automatic operation	1.6	2.6	5.1
2	failure to use low pressure systems when high fail	8.9	9.3	1.6
3	failure to manually operate failed automatic system	2.3	4.0	5.7
4	failure to restore inhouse electrical components	1.4	1.9	3.4
5	failure to restore offsite electrical components	11.2	18.7	5.3
6	failure to perform manual scram	0.1	0.2	4.2
7	failure to manually override automatic system	3.8	7.8	7.2
8	failure to provide SLCS	2.3	5.5	2.4
9	failure to request last resort systems	1.4	20.9	45.9
10	failure to locally operate manually operated component	7.1	13.7	6.6
11	failure to manually override false control signal	10.5	12.1	2.4

sequences; 3 and 4, the station blackout sequences; and 8, a total loss of high pressure water to the core, are risk-significant sequences (as well as many others not studied by Sandia). The specific actions for these sequences are listed with median and mean response times and the error factors for each action. These TRCs are plotted in Figure 5-6. Table 5-5 lists these median and mean response times and error factors.

Compiling the data, as was done in the Sandia report, decreases the calculated uncertainty of the resulting TRCs by producing more numbers per curve, but may mask important characteristics or patterns in the TRCs. The statistical analysis of the data performed by Sandia seemed to show that there were no patterns to lose.

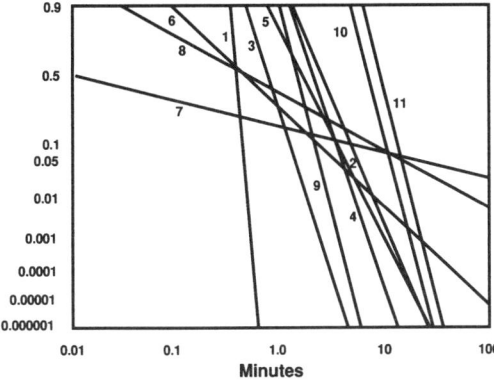

Figure 5-6. Data from LaSalle simulator for selected risk-significant specific actions. (Interpolated from Whitehead, et al., 1987.)

Table 5-5
TRC Parameters for Specific LaSalle Actions

	Drill	Description	Response Time Median (min)	Mean (min)	Error Factor
1.	1/1b	manual scram after ATWS	0.4	0.5	1.3
2.	1	initiate SLCS after high temperature alarm	2.3	5.5	2.4
3.	3	initiate RCIC after station blackout	0.8	1.3	1.6
4.	3	request DG1B repair after station blackout	2.0	4.0	2.0
5.	3	request DG1A repair after station blackout	1.5	3.5	2.3
6.	3	send operator to open FO13 when FO13 fails	0.5	3.8	19
7.	4	depressurize after delayed station blackout	~0	~25.0	∞
8.	4	request diesel fire pumps after station blackout	0.5	14.5	29.0
9.	8	initiate supression pool cooling after reactor trip	1.4	2.1	1.5
10.	8	depressurize after failure of RCIC	7.0	11.0	1.6
11.	8	inject low pressure water after high pressure fails	8.8	14.0	1.6

However, most of the resulting TRCs lie between TRCs (indicated in bold on Figure 5-5) with median response times between 1 and 20 min and with error factors between 3 and 5 (inclusive). This is a "pattern" that has implications relative to the TRC system described in Chapter 10. There are two "outliers", curve 6 and 8 in Figure 5-5. These too are discussed in Chapter 10.

Chapter 6
THERP

As noted in the history chapter, human reliability analysis evolved into a comprehensive discipline in the 1960's. The dominant approach to HRA at that time eventually became known as the Technique for Human Error Rate Prediction (THERP). The early concepts of THERP were used on the landmark PRA, WASH-1400 (USNRC, 1975) as well as most of the NRC-sponsored and many of the industry-sponsored PRAs that followed. Some of the techniques described in subsequent chapters will be but variations on THERP, while others will be logical extensions to ideas not pursued in the basic documentation of THERP (Swain and Guttmann, 1983).

Task Analysis as a Human Reliability Model

THERP arose as a response to a particular type of reliability question:
> in a well specified task, what is the probability that a human practitioner of the task will err enough to lead to failure of the task?

The tasks that initiated THERP involved bomb assembly in a military facility in the southwest US; the failure of such a task could easily have had spectacular consequences. Further, the fact that the task was well-specified meant that significant errors would be, in fact were, rare. So, data was not likely to be available in sufficient quantity to produce the desired probability estimates.

The founder of THERP drew from an analogy to hardware reliability analysis. A task is a minimal set of human actions that accomplishes a specific goal (or mission, in military jargon). The actions are discrete, e.g., open a specific valve or wait until pressure is a certain level. Activities in a task may be performed by an individual or in a team. In a team setting, some actions may be performed in parallel. Some actions may have to precede another action in order for the subsequent step to be carried out, e.g., an operator may need to isolate a safety system pump from its normal flow paths in order to test it. Other actions may come on an as-needed basis, e.g., verification that an action is successful. Basically, however, a task is a serial list of actions.

An example task analysis is presented in Table 6-1. The task is to respond appropriately to a small loss of coolant accident (LOCA) in a PWR. There are four general activities in this response: diagnose the accident, attempt to isolate, i.e., stop or contain, the LOCA, monitor the automatic safety system actuation, and obtain long-term cooling. A PWR has been designed to respond automatically to such an

Table 6-1
Example Task Analysis

GOAL
Obtain reactor core system water makeup and cooling following a small loss of coolant accident (LOCA).

Step	Means
Diagnose event	
detect plant upset condition	several alarms
observe RCS level indicator	pressurizer or reactor level
observe decreasing RCS pressure	pressure indicator
observe sump level increasing	level indicator
observe containment pressure increasing	pressure indicators
observe no secondary side radiation	radiation monitors
Isolate the LOCA	
close PORV block valve 1	one valve control
close PORV block valve 2	another valve control
close letdown line	one valve control
close RCP seal isolation valve 1	one valve control
close RCP seal isolation valve 2	another valve control
Verify Safety System Actuation	
observe HPI pump meters	two flow indicators
start HPI pumps	two pump start controls
observe AFW pump meters	two flow indicators
start AFW pumps	two pump start controls
Obtain Long-term Cooling	
await low level tank alarm	one level indicator
open sump valve 1	one valve control
open sump valve 2	one valve control
open tank valve 1	one valve control
open tank valve 2	one valve control

event and the operators should have no major actions to perform until long-term cooling must be affected. Until then, the operators' role is supervisory, unless safety equipment fail. Even the isolation effort is not necessary but its achievement would lessen the impact of the LOCA. Figure 6-1 is a task flow diagram that depicts the task as four major activities with one of its alternatives (step 5). Any deviation that can not lead to the task goal is an error (steps 6 through 9). Table 6-1 also lists the likely discrete actions that must be performed in order to accomplish each activity; thus, the number of discrete actions (steps) is really on the order of twenty.

A deviation from an intended task step is an error and can arise in one of two ways. The step may be omitted altogether, e.g., a valve may not be opened. This is called an omission error. The step may be performed, but incorrectly, e.g., the valve may be turned the wrong way. This type of error is called a commission error, i.e., the task practitioner commits an action that is inappropriate. Commissions can be further subdivided into selection errors, e.g., the wrong control is manipulated; sequence errors, step n is performed before n-1, when order is important; time errors, the step is performed too early or too late; or qualitative errors, the equipment is manipulated too little or too much. In this way, the THERP error categorization is drawn up independently of the actual human processes or mechanisms that produce the error. Thus, when considering both omission and commission failure modes, the potential number of errors (with their modes) in the LOCA example is at least forty.

A task can be documented in a written procedure, which, in turn, can be used in training on the task and in providing a mnemonic aid when performing the task. THERP provided another use of a procedure, as the basis for the model of the task. If a task were indeed a series of proceduralized steps, then task failure, i.e., task unreliability, would be the logical sum of the failures of each step. This can be expressed probabilistically as:

$$\Pr[\text{task failure}] = \sum_i \Pr[\text{step i failure}] \qquad (1)$$

where $\Pr[x]$ is the probability of x and $\sum_i$ is a sum over the index i.

This formulation is equivalent to the formulation in hardware reliability analysis for a system failure probability given its component failure probabilities, where the failure of any component failure leads to a system failure, i.e., the system has no redundancy. Sources of redundancy in human tasks include monitoring one's own actions (leading to the possibility of self-correction of errors), verification of steps by other people, and instrumented changes in the system that can indicate that an incorrect action has been made. If a step has some potential for redundancy, then its probabilistic formulation is:

62   Human Reliability Analysis

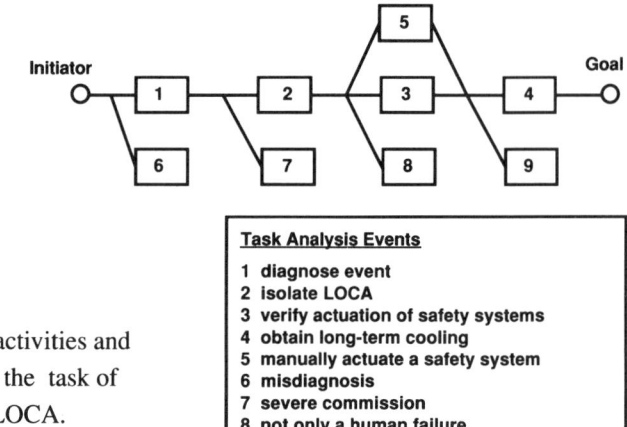

Figure 6-1. Major activities and potential failures in the task of mitigating a small LOCA.

Task Analysis Events
1 diagnose event
2 isolate LOCA
3 verify actuation of safety systems
4 obtain long-term cooling
5 manually actuate a safety system
6 misdiagnosis
7 severe commission
8 not only a human failure
9 no action or action too early or late

Pr[failure in a redundant step] =
   Pr[failure in initial step]
      x Pr[failure of redundant step | initial step failed]   (2)
where Pr[x|y] is the conditional probability of x upon condition y.

This formulation is directly analogous to the failure probability of a two-component hardware system, each of which may provide the same system function. The conditional probability can be simply related to the unconditional probability by:

Pr[x|y] = a + b Pr[x]   (3)
where a and b are some positive numbers.

When a is zero and b is unity, x is said to be independent of condition y. When a is unity and b is zero, then Pr[x|y] is unity and x is said to be completely dependent on the condition y, i.e., x is assured to occur if y occurs. When values for a and b range between unity and zero, this gives rise to degrees of dependency. Note that when, for example, a is zero and b is greater than unity, the dependency is in some sense "positive", i.e., condition y makes it more likely for the opposite of x to occur.

The degrees of (negative) dependency were partitioned in THERP using the values for a and b as in Table 6-2. This set of values amounts to the so-called dependency model in THERP. Note that these values were obtained from judgment, not data.

Table 6-2
THERP Dependency Model

Level of Dependency	a	b	Pr[x | y] Actual	Approximate*
zero	0	1	Pr[x]=P	P
low	0.05	0.95	(1+19P)/20	0.05
moderate	0.14	0.86	(1+6P)/7	0.14
high	0.5	0.5	(1+P)/2	0.5
complete	1	0	1	1

* when P=Pr[x] is small, i.e., less than 0.01.

A similar concept was also used in the military analyses of hardware reliability to adjust "basic" probabilities of component failure by one or more factors to reflect dependencies of the component failure on environment, manufacture, installation, etc. (US DoD, 1982) THERP adopted this approach and called the adjustment factors "performance shaping" factors (PSFs). Almost anything, in THERP's view, can be a PSF.* Chapter 9 further discusses PSFs, or influences as they are called.

Thus, each basic step failure probability in (1) could have its own PSFs associated with it and the conditional probabilities of any redundancies (dependencies). The result is a computational form as follows:

Pr[step failure] = Pr[basic step failure] x α x Pr[failed redundancies]   (4)
where α is the product of the values of one or more PSFs.

Task failure quantification would then amount to first identifying the basic steps that must be performed in the task, then identifying the influences on each step that make the probability of the step something other than nominal, and then identifying any possible redundancies for each step. These basic step failure probabilities, PSFs, and conditional probabilities would need to be generated from some data set. The task failure probability would be calculated using the straightforward algebra of (4) and (1).

---

* Note that the term "PSF" applies both to the influence itself, e.g., the adequacy of procedures, and to its numerical value as a factor in (4).

64   Human Reliability Analysis

Note that, although a task will be performed over time, its analysis into discrete steps all but eliminates this time-dependence. Each Pr[x] can be estimated as would a demand failure (as in Chapter 4), i.e.,

Pr[step failure] = e / n   (5)

where e is the number of errors (omissions or commissions) in a **similar** step and n is the number of times that a **similar** step is attempted.

In this way data can be imported from different industries and applications to support calculations like (5). [The data used in THERP is discussed in the previous chapter.]

The total HRA process of THERP evolved until it could be described, as in Figure 6-2, as consisting of a familiarization effort, a qualitative assessment, a quantitative assessment, and an incorporation effort. The familiarization effort was required to identify the tasks that needed to be quantified, to review their procedures, and to gather information about the plant and the people who operated the plant. The

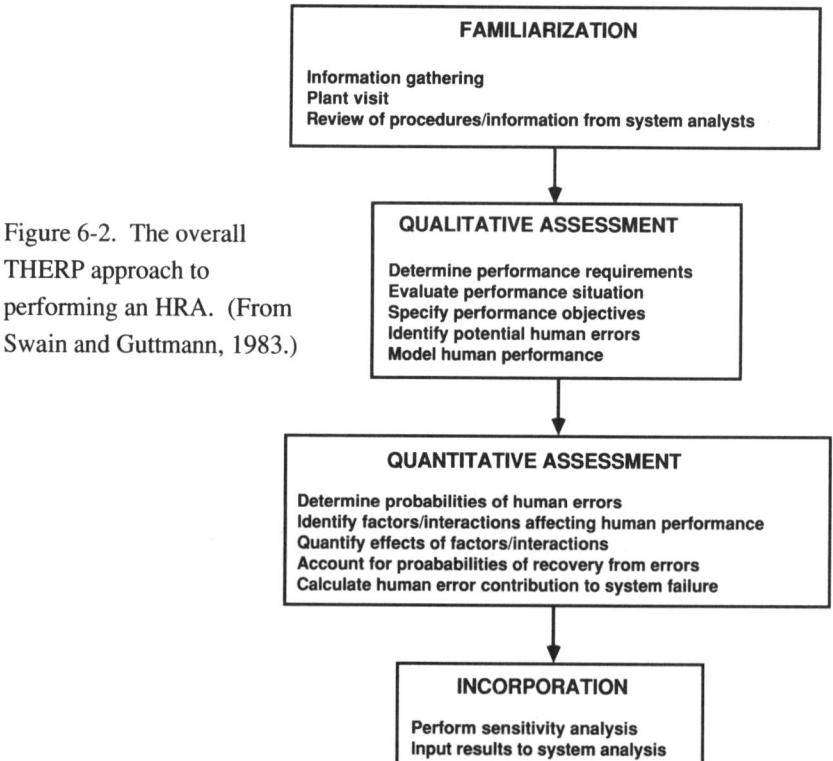

Figure 6-2. The overall THERP approach to performing an HRA. (From Swain and Guttmann, 1983.)

qualitative assessment performed the task analysis, i.e., identified the tasks to be assessed and their task steps, their redundancies, and the PSFs that influence each step. The task was modeled using what was called an HRA event tree. The quantitative assessment applied (1) to each task. An uncertainty estimate was also provided. The incorporation effort provided the PRA analysts with the task failure events and their statistical parameters and assessed their impact on the plant's risk.

The next section discusses the HRA event tree concept and quantifies the example.

The HRA Event Tree

The structure used to "model" the task failure event was eventually called an HRA event tree, to distinguish it from the PRA event trees. An HRA event tree merely captures the essence of Figure 6-1 and Table 6-1, for example, in a "failure logic". Figure 6-3 is an HRA event tree for this example and includes failure probabilities at the end of each failure path taken from the THERP data. Equation (1) allows the

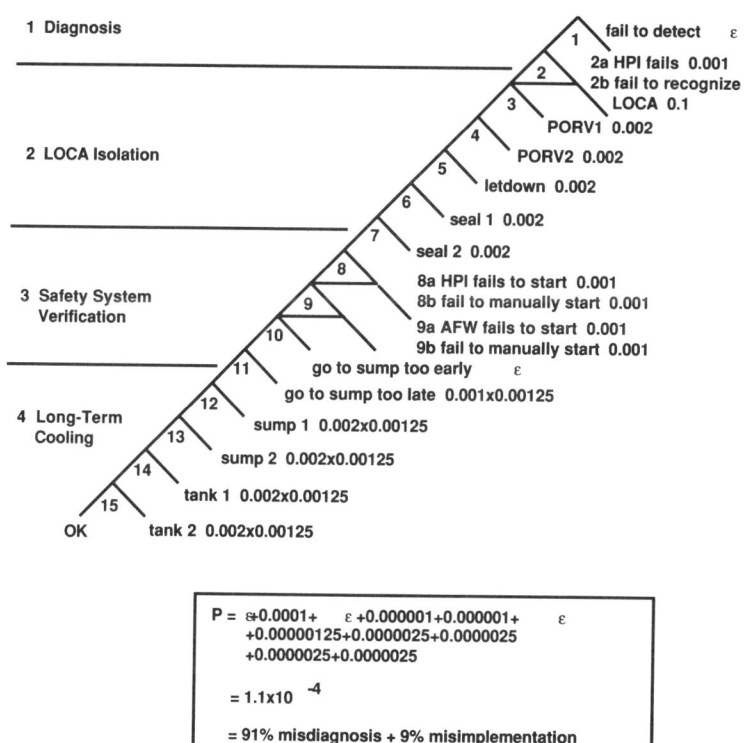

Figure 6-3. An HRA event tree model of the task of mitigating a small LOCA.

66 Human Reliability Analysis

task failure probability to be calculated as 1.1×10^{-4}. About 91% of this failure probability is due to misdiagnosis in the early stages of the event; 9% is due to failures in implementing the late, long-term cooling phase. The fact that diagnosis dominates this task failure is typical for post-initiator events. Other tasks may also be influenced by decision making or other sources of burden.

Several features of the THERP system should be noted:

1. Diagnosis is considered a "holistic" process and thus, has only a single value associated with it. This value depends on the available time and uses the THERP TRC (see Chapter 10).
2. A manning model is implicit in the quantification. It was assumed that one operator would be in the control room at the inception of the event, a second operator would arrive by 2 min., the shift supervisor would arrive by 5 min., and the shift technical advisor would arrive by 15 min. The dependency factors over time were calculated from the dependency model and assumed to be as in Table 6-3.
3. Task failure is integrally involved with hardware equipment failure. Task analysis cannot alone be used to quantify HRA. The early failure and late failure subevents in the HRA event tree would probably be identified as separate events in a PRA structure and be incorporated separately.

Table 6-3
Manning Model in THERP
(From Swain and Guttmann, 1983)

10 minutes
  OP 2 has complete dependency
  SS has high dependency
  joint dependency factor is $1.0 \times 0.5 = 0.5$

20 minutes
  OP 2 has high dependency
  SS has moderate dependency
  STA has high dependency
  joint dependency factor is $0.5 \times 0.14 \times 0.5 = 0.035$

30 minutes
  OP 2 has high dependency
  SS has low dependency
  STA has low dependency
  joint dependency factor is $0.5 \times 0.05 \times 0.05 = 0.00125$

Extending THERP

The THERP technology is a rich compendium of HRA source material developed over two decades. THERP, however, has some limitations. Some are deliberate omissions and others are limitations that arise in the practical application of the HRA process defined by THERP. This approach presented in Part 2 will go "beyond" THERP in the sense of addressing these self-imposed and practical limitations. As an indication of what will follow, the major limitations that need new technology are:

1. Diagnosis is accounted for in THERP by adjusting a 0.1 screening probability by a dependency factor based on the manning model. This nominal diagnosis model also does not account for the new functional symptom-oriented emergency procedures, developed for all LWRs since TMI.
2. Decision making in THERP is accounted for quantitatively by assuming decision failure modes to be contained in the uncertainties of the basic estimates, in particular the diagnosis estimates. Conflict, hesitancy, confusion, and other negative influences on decision making are part of the case histories in all high-technology, high-risk arenas and must be addressed explicitly. Burdens of all types must be addressed as much as possible.
3. The subjectivity of the THERP use of PSFs has been heavily criticized. Response to this has gone two separate directions: toward objective criteria, such as TRCs developed from simulator experiments; and toward to fully subjective systems, such as SLIM (Embrey, 1983). Neither of these options seem correct alone. As much objectivity as possible is desirable but is not yet fully feasible.
4. The tasks of an HRA process indeed should include those in Figure 6-2. However, in the past, HRA, using THERP, was often practiced in a vacuum relative to the overall PRA analysis. This led to problems in integrating the HRA within the PRA, its customer. Events that were assessed because they were interesting from a human reliability point of view might turn out to be risk insignificant, while events that were risk significant might not event be analyzed. Economy and focus has shown the fourth task area—incorporation—to be of much more importance than granted in THERP.

The next chapters describe an HRA process that addresses these issues.

PART 2
An Integrated Approach

Chapter 7
HRA AS PART OF RISK ANALYSIS

Risk is a chance that some event will occur. Since the term "risk" is used when the events of concern can lead to harm or loss rather than benefit, risk is directly associated with events that are referred to as accidents, while the potential harm or loss at risk is called a hazard. Risk analysis identifies, characterizes, and (sometimes) quantifies the risks from a technology or a natural phenomenon.

There are several ways to perform a risk analysis, but they can be categorized into two general types: the analysis of actuarial data and structured analysis. The risks people face from automobiles, commercial airflight, or cancer, for example, are usually computed from actuarial data, since such data are sufficient to make estimates of these risks with reasonable confidence. Nuclear plant risks, however, are so rare that direct data analysis is not possible. No one has been killed or injured offsite from a US nuclear plant accident and Chernobyl is the only lethal accident due to a failure to contain radioactive materials in a nuclear power plant. There have been several fatalities related to radioactivity in test reactors around the world. Also apparently, a boy was killed in an automobile accident during the evacuation of the area around TMI.

The alternative to data analysis is to decompose the risk of interest into smaller elements until data is sufficient enough to quantify the pieces. These estimates are then propagated back through the structure to produce an overall estimate of the risk. Several structural risk analysis techniques exist—the Management Oversight and Risk Tree method (MORT-Johnson, 1980), cause-consequence diagrams (Nielsen, 1975), hazards and operational analysis (HAZOP-Lees, 1980), and probabilistic risk assessment (PRA, see the Appendix for details). The HRA process described in the next chapter focuses on its application to PRA but is general enough to be adopted to any structural risk analysis approach.

A great deal of uncertainty can be associated with assessing risks. These uncertainties arise from lack of knowledge of the phenomena under investigation, of the occurrence frequency of some events, and of the performance of both people and equipment involved. The performance of a risk analysis reduces the uncertainty concerning some of the elements of risk so that attention can be better allocated to those areas that are risk relevant rather than all conceivable hazards. In this way, the performance of a risk analysis of and by itself can reduce risk, i.e., knowledge is gained even without making changes in design or procedure, because it can better focus the awareness of the organization responsible for the technology on those areas of risk significance.

A risk analysis involves many elements, as depicted in Figure 7-1. The major groups are (1) a model of the events that can lead to a specified but general risk, e.g., risk from different kinds of fire, (2) models of the consequences or effects of the events, e.g., property damage from various kinds of fires, and (3) techniques and data used to understand and quantify the risk in probabilistic terms.

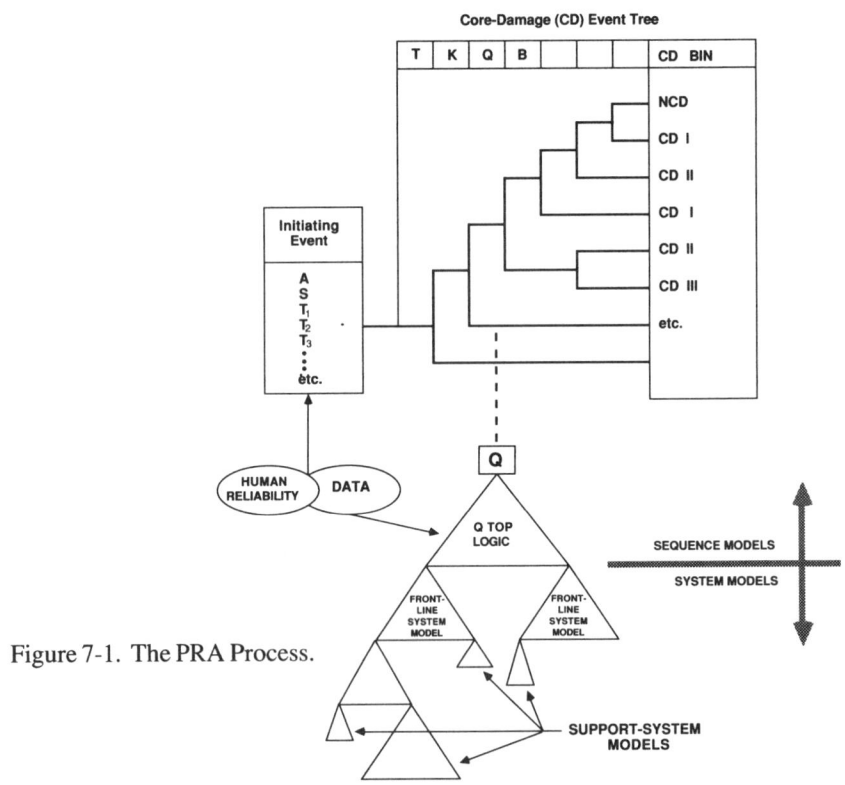

Figure 7-1. The PRA Process.

Human reliability analysis can support a risk analysis in each of the levels depicted. Human events can contribute to the loss of productivity, the initiation or prevention of damage states, and to the mitigation of the consequences of these states (see Figure 7-2). Human interaction affects the plant as one of three general types—failures in test and maintenance activities, failures in operating equipment (e.g., acting as the actuator of a valve), or failures in backing up failed equipment or finding recovery alternatives to the failed equipment. The first type of failure often leaves equipment unavailable but has no other effects. These types of failures are called "latent" and may affect production equipment or equipment reserved for incident and emergency management. Failures in the role of actuator usually has immediate effects, the equipment may fail or the reactor may trip, causing a loss of productivity. There have been no known incidents in the nuclear industry of a single operator failure leading to a loss of incident or emergency management capability. Finally, whenever equipment does not operate as intended, people may try to restore its functioning or find alternative recovery options. In studying human interactions, HRA is, thus, integral to a full risk analysis.

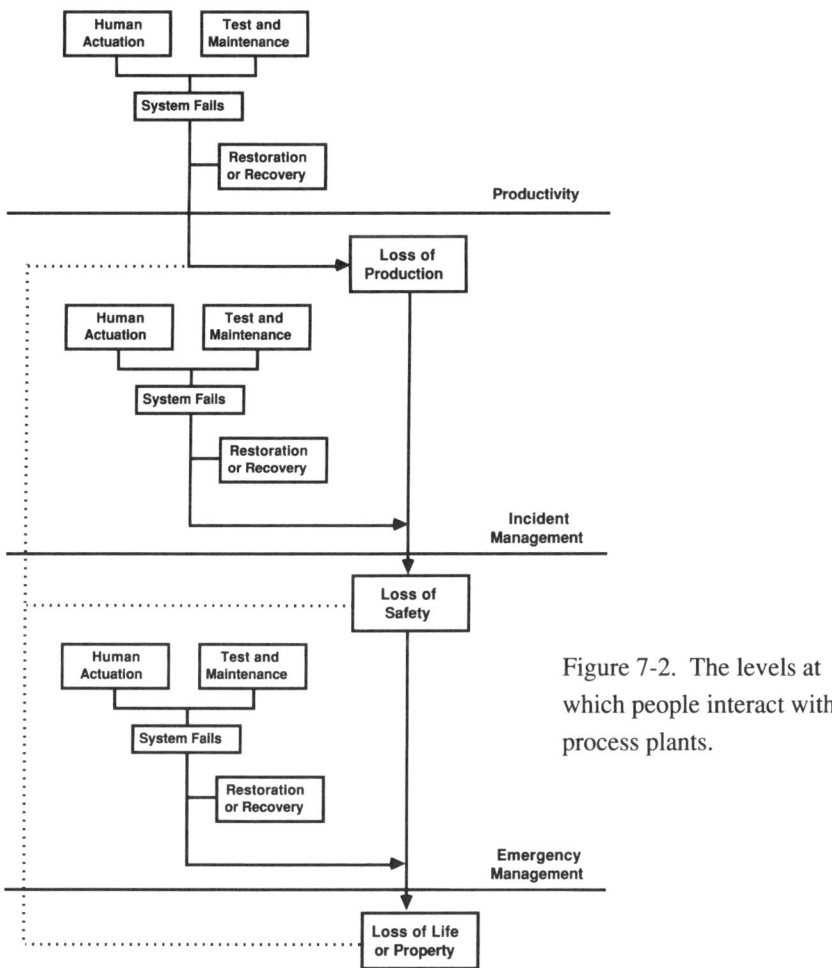

Figure 7-2. The levels at which people interact with process plants.

The variation on HRA needed to support a risk analysis is a subset of HRA methods, and in turn, a part of the general class of human factors techniques, that includes task analysis, link analysis, and detailed control room design review. When integrated into a risk analysis, HRA is also a type of system reliability analysis technique, which is used to model the plant and its systems.

The customer for HRA is the larger risk analysis. This supporting role constrains the methods that can be effectively used in an HRA; but the peculiar requirements of HRA also affect the models of the risk analysis. In its risk context, the goals of a full HRA consist of:

1. Representing the plant's risk contribution due to its people and their supporting materials, such as procedures,

2. Providing a basis from which plant managers may make modifications to the plant while optimizing risk reduction and enhancing human factors,

3. Assisting the training of plant operators and maintenance personnel, particularly in contingencies, emergency response, and risk prevention.

The HRA procedure discussed in the next chapters is adapted to a particular type of risk analysis referred to as a probabilistic risk assessment (PRA). There are many PRA structures that could be presented; the one that was chosen was borrowed from the PRA of Oconee Nuclear Station, Unit 3 (NSAC, 1984). This choice was made because this structure is reasonably representative of the way PRAs are performed and because it has been published in a manner that is fairly accessible.

This chapter briefly describes the PRA process and its relationship to the HRA. A more detailed discussion is contained in the Appendix.

Overview of PRA

A PRA is an engineering analysis of the possible contributors to the risk from a nuclear power plant (or some other technology or natural phenomenon). This risk can be measured in terms of the postulated number of fatalities from a potential release of radioactive materials, the potential for economic losses from a major accident, or the estimated frequency of major damage to the reactor core or peripheral equipment. The results from a PRA can be of three main kinds:

1. A description of equipment failures, human failures, and process events whose combination must occur before a specified hazard can occur,

2. A quantitative estimate of the risk in one of the above terms, with some indication of the uncertainty of this estimate, or

3. Relative quantitative measures of the risk-importance of equipment, human factors, plant design parameters and policies, etc.

The key to understanding the risk of a plant is the identification of potential sources of hazards and the tracing of the sequences of events that can lead to the realization of these hazards. The characterization of these sequences forms the framework for the analysis and results of the PRA.

The PRA of a nuclear generating station is a significant enterprise that may deal with many, diverse disciplines, such as reliability analysis, thermal hydraulics analysis, core physics, and environmental transport. To make the process manageable, a top-down approach has been developed in the nuclear PRA community (see USNRC, 1983, for example).

First, a risk criterion and its associated hazard are specified, such as the frequency (criterion) of core damage (hazard). Then accident sequences that would lead to the specified hazard are characterized functionally and chronologically in general terms. An event tree or other logic diagramming technique is used to systematically delineate these sequences. The resulting sequence descriptions include the event that initiates the accident — the initiator — and any other events that must occur prior to obtaining the hazard.

This top level description of accident sequences is usually too general for the estimation of sequence frequencies or other risk quantities directly from historical data. Therefore to extend whatever data is available, the events in each sequence are further modeled using techniques from system reliability analysis. The system models are usually developed as fault trees (Vesely, et al., 1981), but may be, alternatively, GO models (EPRI, 1983) or reliability block diagrams (IEEE, 1975). This modeling may involve several levels in order to assure that all equipment that can play a significant role in a sequence is properly included.

Sequence descriptions are generated from the system modeling using computers to solve the models. These models have a mathematical representation as Boolean equations for a specified, undesired event, such as a system failure, called a "top event". The Boolean equation is expressed in terms of elemental events, the "leaves" of a fault tree. These elemental events are called "basic events" and sets of basic events that would cause the occurrence of the top event were they to occur are called "cutsets" (since together the events cut success paths). This list of cutset descriptions of the sequences can be quantified by providing estimates of the failure rates and other reliability parameters, such as demand failure probabilities and unavailabilities, of the basic events in the models, typically equipment failures or human failures. The sequences then can be prioritized by total sequence frequency and aggregated into "bins" of sequences that have similar phenomenal characteristics. Thermal hydraulics (T/H) analyses are used to determine the success criteria of systems, such as how many pump trains must operate to supply sufficient flow for system success. It is these criteria that must be postulated to fail in a sequence for cutset to occur. T/H analyses are also used to specify the phenomena that are used to aggregate the great number of possible undesired phenomena into a manageable number of "bins". If the risk criterion requires it, the accident sequence is then developed from core damage through containment phenomena to the release and transport of radioactive materials. The additional sequence details are quantified using equipment failure data or engineering judgment.

The result is a spectrum of consequences with associated frequencies and their uncertainties. This spectrum can be compiled into an aggregate estimate of risk, such as a point estimate of the core damage frequency, or expressed as a distribution, such as the complementary cumulative distribution function of early fatalities.

The diagrams in Figure 7-1 represent the way in which certain sequences of events are identified that can lead to specific damage states. System reliability models are used to logically decompose and analyze the identified sequences. Together, these diagrams represent a model of the events and their consequences.

The model is then quantified and a measure of the risks are presented as an overall risk curve which has consequences decreasing in probability of occurrence. This quantification requires data and techniques to combine data or estimate quantities for which there are insufficient data. The result is a probabilistic risk assessment.

Human Failures and Accidents

Technological risk is directly associated with off-normal events, often referred to as accidents. The human factor enters into risk as depicted in Figure 7-3. An **accident** is an event or occurrence that has results, effects, or consequences that are bad by some human standard, typically damage to the system or people around

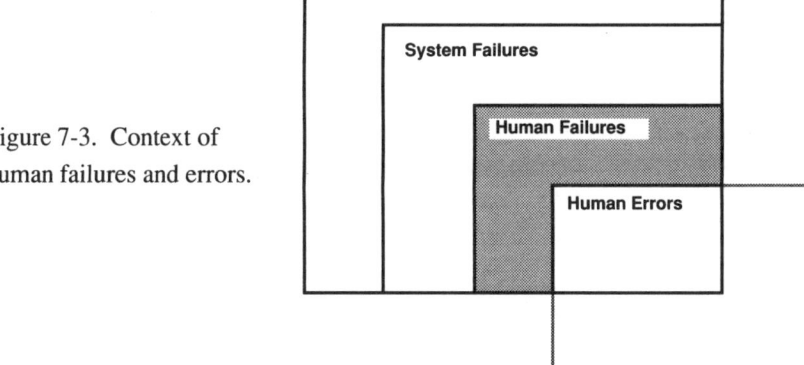

Figure 7-3. Context of human failures and errors.

it. Life is lost, injury incurred, or property damage is suffered. Many accidents are induced by failures of technological systems, which seem to arise from the complexity of the systems themselves (Perrow, 1984). The hardware portion of a **system** is a collection of equipment, i.e., hardware, that has specific, intended functions, and interacts among its pieces and with the people and software that operate the system. A **system failure**, then, is an event or occurrence that has bad consequences upon the system's functioning. Obviously, some accidents, often called catastrophes or natural disasters, occur as the result of events in nature, not from the failures of equipment, software, or people.

Some of the failures of systems are attributable to people. It is common practice to call all such human-induced failures human errors. But everyone has personal experiences with trying to use some gadget that is so ineptly designed in some feature as to practically prevent its successful use. In these cases, the user makes the errors, but the designer is really at fault. To the degree that a failure is attributable to the person who "had it last", i.e., the cause of the error is predominately within the person, then it is that person's error. The problem is that the boundary between a person and the environment is fuzzy and complex. Also, as systems increase in complexity, such as in a nuclear power plant, operator comprehension is impaired—accidents hence are in great part incomprehensible system interactions (Perrow, 1984). The use of the pejorative term "error" is unwarranted in many cases of high-risk accidents and can lead to misallocation of resources and a failure to avoid future accidents.

As a result, the term human failure seems more appropriate in the context of accidents and system failures. A **human failure** is a system failure that can be proximally attributed to the actions or inaction of one or more people. This leaves a **human error** as a system failure that at root cause can be attributed solely to the intrinsic mechanisms of a single person. This includes errors made by people acting in groups, but "group dynamics" leads to human **failures**. In high-risk technologies, human error may not even comprise half of the population of human failures. Of course, many human errors have no system or significant consequences and are not part of an accident's causal chain.

HRA as described in the subsequent chapters identifies, characterizes, and quantifies **human failures** as described in this manner in a PRA.

Incorporating HRA into PRA

The task of HRA in a PRA is as credibly as to possible identify all human failures that can contribute to risk-important sequences identified by the PRA. The event specification must be in terms that make sense in the sequence context, that make sense from a human factors and HRA perspective, and which can be assigned a probabilistic value commensurate with estimating the sequence frequency.

From a logical perspective, human failure events can be incorporated into the PRA models at any place. The computer solution is impervious to the meaning of the event and the computer merely manipulates the event's name as a logic symbol. However, the location of human events in PRA models is one of the more controversial areas in HRA.

The first nuclear plant PRA (USNRC, 1975) associated human failures with component failures, e.g., "operator leaves recirculation valve open", or with major procedure steps, e.g., "operator fails to start the feedwater pump". The assumption of this technique (THERP - Swain and Guttmann, 1983) was that human failures could be decomposed into behavioral units that were directly related with plant equipment and were otherwise contextless. The result in the PRA often was that there were as many human events as there was equipment. The tradeoff, however, was that these events could be quantified generically, e.g., there was a failure probability that applied to **all** failures to open **any** motor-operated valve.

This technique, however, clearly contradicts recent findings related to human error (Reason, 1983) and does not account for the phenomena of cognition (see Chapter 3). Subsequent techniques in HRA assume a more holistic perspective on errors, particularly mistakes. The result on the PRA models is that fewer but more significant human failure events are identified. Further, the events identified are less often scattered about the lower level modules, and are less often contextless, but are more often directly tied to the development of specific accident sequences.

The problem of incorporating an HRA into the PRA was recognized by the partisans of THERP only after conversations at the Myrtle Beach conference on human factors (*Conference Record*, 1982). One visible result was the inclusion of incorporation as an element of the HRA (see Figure 6-2). The assumptions made concerning the causes of human failures directly affect the process by which human failure events are identified and the procedure by which they are incorporated into the relevant PRA models. How these assumptions and the HRA procedure are interrelated is indicated in the subsequent chapters.

Chapter 8
THE HRA PROCEDURE

There is a natural tendency to try to proceduralize an activity in order to stabilize its use and warrant its results. This is why process plants have test, maintenance, and operational procedures. This is also why there have been efforts to proceduralize HRA. This chapter presents one such "procedure" for performing HRA in conjunction with a PRA. As noted in the introduction, the procedure works as measured by the standard of the number of PRAs in which it has been applied. First, an overview is presented and then the three major HRA task groups are described.

An Overview

As can be inferred from the Chapter 7 discussion, in accordance with the systems engineering process of requirements allocation, PRA places certain requirements on HRA. To satisfy these requirements, the HRA must provide assurance that:

1. all risk significant human events are identified,
2. these events can be incorporated into the PRA models, and
3. these events are quantifiable with probabilities.

Each PRA/HRA has attempted to meet these requirements, but often in unsystematic or peculiar ways. However an HRA is performed, there are certain steps that have to be performed, even if not in the indicated order, or the HRA cannot be considered to have been satisfactorily performed. Figure 8-1 depicts these steps. The systematic process portrayed in this figure is similar to a proposed IEEE guide on HRA (IEEE, 1987) and much like that developed by the Electric Power Research Institute in their SHARP effort (Hannaman et al., 1984) and, to a lesser degree, that offered in the *PRA Procedures Guide* (USNRC, 1982—see also Figure 6-2).

A properly engineered PRA approach is characterized by being integrated and economical. Correspondingly, the HRA tasks are partitioned so to match the PRA needs. This results in three groups of tasks: (1) integration tasks, (2) tasks that support the initial quantification(s), and (3) tasks that provide the detailed event analyses at the end of the PRA. Each of these tasks groups are discussed in the following sections.

Integration Tasks

Following the systems engineering tradition, the HRA must be integrally merged with the PRA in order that the overall risk results are reasonable from both the

80   Human Reliability Analysis

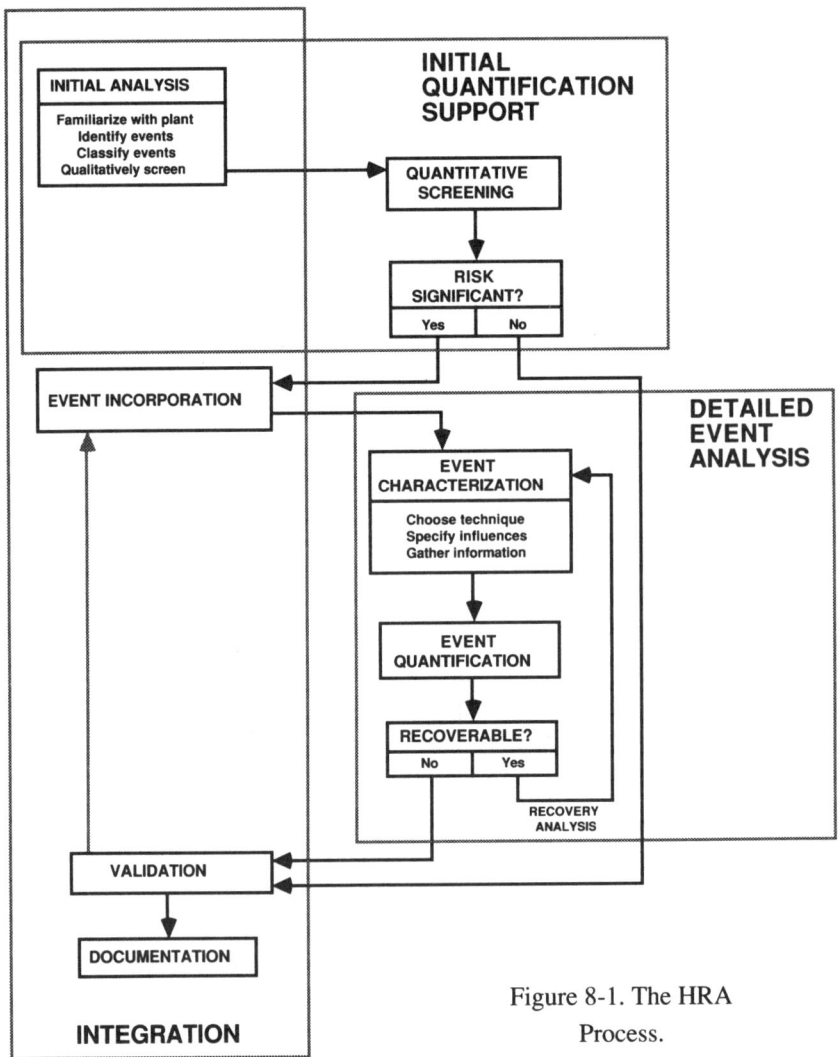

Figure 8-1. The HRA Process.

hardware and human perspectives. Integrating HRA within PRA is no trivial matter. Because integration was not given mention in the early guidance documents concerning HRA, such as the *PRA Procedures Guide* or NUREG/CR-1278, serious problems were created in properly accounting for the impact of human failures in the final results using these documents (this was the so-called incorporation problem mentioned in Chapters 6 and 7). Generally, the integration step is performed best by someone who has PRA experience, particularly by someone with system analysis experience, and HRA expertise. When a single individual is not available with this rather extensive background (as is often the case), a team approach is recommended. When the team alternative is chosen, it is important that all experience areas are represented. It is also important that the team leader be the team member with the most extensive knowledge of the PRA requirements, so that the HRA effort can be focused toward the risk significant areas and not waste effort on incidental but technically interesting human factors' concerns*. PRA considerations lead to the identification of human events and HRA considerations lead to adjustments in PRA models in an interactive, iterative manner. This integration task continues throughout the PRA but the early efforts can result in cost savings and more efficient use of the project resources. The integration effort begins with the analysts becoming thoroughly familiar with the plant, its systems, processes, and activities. The initial analysis should include at least one plant visit to survey the control room or simulator (if there is one), other major facilities, major equipment, and the training facilities. Contacts should be established with the operations, maintenance, training, and procedure development staffs. The plant procedure system, particularly emergency procedures, should be reviewed. A quick review of the control room layout and its key instrumentation and controls should be conducted, taking care to note differences from other plant experiences. This visit will significantly influence the detailed analysis to come and with the visit, the validation process for the HRA begins.

The initial HRA analysis is best performed coincidentally with the initial system analysis effort. Human failure events are identified as the PRA models evolve and are at first specified functionally, i.e., without breakdown into human failure modes and mechanisms. Event identification is the key interface between the system analysis and the HRA efforts. It is clear that there is no way to predict, much less assess, all human-caused events that could have some impact on risk. There are recorded instances of operators sitting on equipment controls and inadvertently actuating them and the mere dropping of an annunciator light bulb initiated a serious incident at the Rancho Seco nuclear plant (Chexal and Wycoff, 1980). However, it is possible to classify human events, with the help of a taxonomy such as Embrey's, and then roughly "account for" the classes. A classification scheme that was developed specifically to assist the HRA integration effort is presented in the next chapter. In addition, the system analysts should be provided

---

* The problem of a lack of "risk focus" has been a complaint often heard from operators whose control room is subject to a human factors review. After the review, they are left with hundreds of so-called human engineering deficiencies (HEDs) but without some risk focus, they have no means to determine which are risk- and cost-beneficial to correct.

with guidance concerning the purposes and information required for the HRA; some of these basic guidelines are provided in the next chapter. The use of the human failure classification scheme and the HRA guidelines does not guarantee that every risk significant human failure will be identified but it does provide a substantial assurance that this will be the case.

The approach recommended is that the human reliability analyst be familiar enough with system modeling to make the HRA and fault tree/event tree development an integrated effort. A crucial reason for this requirement is that—based on the classification scheme and the knowledge of the analysts—many potential human failure events need not be modeled. This selective process is often referred to as qualitative screening, and it can only be done properly if the resources of both types of analysts are directed to the problem.

The integration effort involves another selective activity. The HRA models and techniques that are needed to assess and quantify the human events identified are assigned to each of these events as they are identified. The manner in which a human failure event is identified will depend on how it is to be incorporated into the PRA models. For example, the approach taken in WASH-1400 was to model human events at the level of the component affected by the event. This led to the fractured, task analysis approach characteristic of early THERP models. Later PRAs have incorporated global diagnostic events higher in the PRA logic models. This results in a more holistic approach and is the primary reason why quantification is often performed using time reliability correlations (TRCs) in these later studies. The integration effort, thus, matches the PRA with a particular philosophy of HRA.

As the PRA risk sequences unfold and begin to be understood, the integration effort, by assuring that proper consideration has been given (from both the system analysis and human reliability analysis perspectives) to either including or eliminating human failures, provides some measure of validation of the HRA elements of the sequences. This, of course, cannot be considered to be a complete validation effort, as may be performed in computer software validation and verification, but taking an integrated approach simply means that the sequences are reviewed for their face validity, i.e., whether they make sense operationally. Further validation is obtained in diverse ways. Typically, one or more plant operations people serve as advisors on the PRA team and their experience base provides additional assurance of the validity of the events. The final project review includes the HRA results, which takes advantage of the entire project team's experience. And finally, the sequences are reviewed with operators using the plant simulator, if one exists, or by walkthrough or talkthrough exercises.

The final step of the integration tasks is the development of the appropriate task documentation. Documentation development is an ongoing process that terminates with the presentation of the project report section on the HRA and the HRA workbooks. The HRA process has two potential stopping rules: (1) when no more risk significant human events can be identified, or (2) when no more risk significant human events have credible recovery options. At this point, if not earlier, the whole analysis must be thoroughly documented including, at a minimum, the events identified, the assumptions upon which their analysis was based, and their quantitative results. To aid in this often tedious documentation process, a software system (ORCA, see Chapter 11) has been developed. This code greatly aids in the documentation of the HRA.

Initial Quantification Support Tasks

All human events that are included in the PRA structures are screened, i.e., assigned a conservative estimate for their probability of occurrence. This screening process is part of the first quantification stages of the PRA's system analysis and is used solely to constrain the more detailed PRA analysis to only those sequences of potentially significant risk. Slips are screened based on general patterns from the THERP process as described in Chapter 11. Mistakes are screened by using the time reliability correlations, by choosing the five minute value for the appropriate correlation (see Chapters 10 and 11).

All human events that survive screening, i.e., are part of candidate dominant accident sequences, are then considered risk significant and analyzed in detail. Recovery events are usually adjoined to risk significant sequences, whenever recovery is a credible possibility, and after solving for the sequence cutsets, are also analyzed in detail.

Detailed Event Analysis Tasks

Once a human failure event has been identified, screened, and found risk significant, information is gathered on the situation in which the event is postulated to occur. A specification of those direct influences on the event's reliability is made using human factors techniques, systematic judgment techniques, or by applying procedures described in various quantification techniques. The human failure event is then further decomposed into its credible failure modes and represented in a logic structure.

The technique that will best apply to the relevant influences and failure mechanisms is then chosen, as discussed in subsequent chapters. Actual application of the technique is usually straight-forward and is shown by examples in subsequent chapters. When the event is characterized in enough detail to apply a quantification technique, a computer code called ORCA allows the quantification to be performed automatically.

A sequence that is recoverable is of potentially less risk than a similar one that offers no opportunity for recovery. A recovery human failure event is the failure to restore failed equipment or to find alternative equipment or configurations within the time period required. Recovery events typically are not identified or quantified until sequence specifics, i.e., cutsets, are known. This is an economizing strategy, since it defers detailed analysis of many events until after the bulk computer solutions are made. If a sequence is recoverable, then its recovery analysis is an iteration through much of the PRA/HRA process, including system as well as human modeling and quantification. The detailed analysis effort is always required. The approach presented here considers recovery activities even if they are not supported directly by procedures but by plant training or operator experience. The recovery analysis is also the area in which HRA can credit the plant operators for emergency preparedness and other emergency response activities that have evolved post-TMI in the nuclear industry.

The next chapter discusses the qualitative analysis performed in an HRA. Chapter 10 develops the TRC system incremental to quantifying post-initiator human failure events. Finally, Chapter 11 describes the quantification techniques used to screen human failure events, and quantify slips and mistakes in detail.

Chapter 9
QUALITATIVE ANALYSIS

As Chapters 10 and 11 will demonstrate, the quantification of a human failure event is fairly well-structured, *once* it is identified and characterized properly. The qualitative analysis of an HRA is where much of the controversy and most of the effort reside. The qualitative analysis of an HRA in a PRA consists of four basic tasks—familiarizing with the plant, its procedures, policies, operators, instrumentation and controls, systems, and the PRA models; identifying the risk-significant human failure events; incorporating the events into the PRA model structures; and characterizing the events in such a way that all major influences on their reliability are accommodated.

A human failure event can thus be thought of as a "vector" consisting of three sets of information:

1. The mechanistic characterization of the event, i.e., the failure type, its mode(s), the likeliest mechanisms of failure, and the potential causes. This description of the event follows the taxonomy of Table 3-2.

2. The event interface specification, i.e., the fault tree(s) or event tree(s) that the event is applicable to and the logic and gate identification of the incorporation.

3. The identification of the major influences on the likelihood of the event and an assessment of the importance of each influence.

The identification of human failure events is performed under the auspices of an event/failure mode classification scheme. Incorporation of human failure events into the PRA is typically a joint effort between the human reliability and systems analysts and some guidelines on this process are provided. The last section of the chapter describes the ways in which the major influences on the reliability of the event are identified. First, the familiarization process is outlined.

Plant Familiarization

An ideal practitioner of a specific plant's HRA would know the plant as the operators, maintenance people, designers, and managers do, all at once. Therefore, the first and continuing effort of an HRA is to approximate this ideal as much as resources allow. Sources of information needed to familiarize with the plant include instrument and piping diagrams, system descriptions, plant layout drawings, normal, abnormal, and emergency operating procedures, maintenance procedures, and work order and control room logs.

Plant presence is a necessary ingredient in HRA familiarization. Visits to the plant must include an initial plant tour, a survey of the control room or simulator (if there is one), and may include visits to specific equipment or facilities as the candidate risk-significant sequences are identified in the PRA. The more the number of visits, the better the information transfer from the plant to the HRA; all within the resources available.

A crucial input to an HRA are the operators and support people. Information can be obtained from them in interviews, walkthroughs of specific sequences, or indirectly by observing simulator exercises when a plant has a simulator. The simulator is probably the most informative medium relative to post-initiator events. It is also highly recommended that an operator or former operator should be part of the HRA team to serve as liaison to the plant and a critical reviewer of the HRA.

The candidate risk-significant sequences should always be discussed with operators in at least a talkthrough process. A photo survey is a good tool to record walkthroughs but permission to conduct such a survey in a control room is difficult to obtain in US plants. Chapter 13 shows a few photographs taken from the photo survey used in the Garoña HRA. The primary use for the photo survey, besides documentation, is identify and document key instruments and controls, key decision points in a sequence, and a general confirmation of the event identification process. More extensive photo surveys of the control room and peripheral facilities can also be used to support the analysis of influences on the human failure events.

The human reliability analysts should also review the PRA logic models—the event trees, fault trees, and initiator models—for their information relevant to identifying human failure events and the interface to the HRA. This activity is not documented as part of the THERP approach and its omission often led to inconsistencies in the other PRA activities and the HRA. This problem is the reason that HRA is treated in this approach as an integral part of the system analysis.

These are briefly the main elements of the familiarization effort in an HRA, which are not easily described in an itemized list of activities. The amount of effort devoted to each element is dependent on the analysts' experience base. However, the familiarization process may require as much as half of the available resources for an HRA and is the foundation for a credible analysis.

Classification of Human Failures

The principal strategy in reliability analysis is to decompose failure events into meaningful subevents for which there are sufficiently available data to make an estimate of some reliability characteristic. The purpose of classifying failure events is to uncover these reliability-related subevents (such as individually replaceable components) and to form the basis for collecting or understanding data concerning these identified subevents. This classification process should be based on physical principles, e.g., failure analysis, or at least observation. The more realistic the foundation for the classification scheme, the greater the predictive power and credibility that the classification can claim. HRA also adopts this classification strategy.

An early classification scheme for human errors, used in THERP, was to categorize all human errors (failure events) as either commissions or omissions. This classification assumed that the reliability of tasks could be assessed by decomposing them into serially performed subtasks much in the same way that routine task performance can be institutionalized by writing step-by-step procedures. This approach assumes that every task must be performed and performed correctly for success. If a step was left out, an error of omission was said to have occurred, and if a step was otherwise performed incorrectly, an error of commission occurred.

Although this scheme partitions errors, it has no direct relevance to the mechanisms or causes of errors. For example, a person could just as easily commit an error as to omit a step because of, say, a lapse in attention (Reason, 1983). As a result, there would be no reason to expect a difference in any reliability parameter based on this distinction. Another weakness of this scheme is that it requires the consideration of any error in any subtask because it may represent a task error without any relevance to risk. For example, the failure to verify an automatic system's actuation is an error related to the objective of assuring system operation but it is irrelevant in a risk sequence that does not also assume the failure of the automatic actuation.

Several other ways of classifying human failures have been devised and the more frequently used ways are described next.

Failure Modes Classification

Rather than looking for the effect that a human failure has on the system that the person interacts with, an alternative, used in most PRA/HRA studies recently, is to look at the mechanisms of failure as a basis for classification. These mechanisms can be grouped into general categories, call modes. These two modes are mistakes and slips (Norman, 1983), where a mistake is a cognitive error and a slip is a subcognitive error. The motivation for this classification scheme is operational. Since slips, or their behavioral basis, are more mechanized, i.e., routinized,

88   Human Reliability Analysis

practiced, and subconscious, they should have different reliability characteristics from mistakes. Since slips usually occur in a less time-constrained environment, it is reasonable to believe that their associated error probabilities are less time dependent, more structural, and arise as faults in an intended protocol or procedure. On the other hand, mistakes occur more often in a time-constrained environment and are more time dependent (it takes "process" time to think up an appropriate response in a novel situation). Mistakes are less structural, and often have little to do with protocol or procedure.

This distinction, while reasonable, is admittedly speculative; however, it has a robustness, exhibited in Table 3-2, not possible in the THERP classification.

Sequence Phase Classification

PRA also induces taxonomic considerations on HRA. A PRA considers two main time phases: the time prior to the initiator and the time after the initiator. An operator's role also changes over these time periods: from normal maintenance and operation activities, which are directed by familiar procedures, to incident investigation and management, and then, infrequently, to emergency response (see Figure 7-2). These latter activities are driven by the event at hand and its perceived severity.

Human failures can be thus divided so as to relate to the phases of an accident (IEEE, 1987):

1. Failures in planned (normal) activities, i.e., so-called pre-initiator events,

2. Failures in planned activities that lead to an unplanned reactor shutdown, i.e., human-induced initiators, and

3. Failures in event-driven (off-normal) activities, i.e., post-initiator events.

The first two categories are behaviorally identical, differing only in whether the results of the failure immediately upset plant phenomena or not. The first category of plan-driven failures are termed "latent" because such errors leave equipment in an undiscovered, inoperable but as yet unneeded state. Category 2 failures directly induce an off-normal condition and, as such, are human-induced initiators. Activities required for incident management are often rule-based because most of the effort in designing off-normal procedures focuses on "anticipated" process conditions about which knowledge can be extrapolated and a response formalized. In accident conditions, however, it is highly likely that the actual conditions will differ from the anticipated conditions enough to require on-the-spot responses in a time-constrained environment. These conditions are really severe and seldom arise, but cannot be ignored in a PRA, require the restoration of equipment or the finding of alternatives among resources. These latter activities are often referred to as "recovery actions". Severe accidents are typically too unpredictable or unlikely for much effort to be expended in proceduralizing an extensive response to them.

Procedure Based Classification

The degree of preparation expended on an activity is another categorizing factor. Some plan-driven activity is according to a specific procedure for an established task. Other normal activity is more random or spur-of-the-moment. Tasks are not specifiable in this case.

Event-driven activity, on the other hand, may also be anticipated by procedure, e.g., the new so-called symptom-based emergency procedures. Failure in this activity can be referred to as response failures, in that a human response was anticipated and designed for. Recovery is more ad-hoc and less directly supported by procedures.

Each of the resulting four event types could include either slips or mistakes as error types. Unspecified, normal activity cannot be decomposed any further. Proceduralized actions in routine task activities are prescribed by supervision or work order and there is little opportunity for mistakes; thus, slips dominate. During off-normal conditions, awareness is heightened, more than a single person is around, while time is usually not as forgiving. Thus, slips will not often go unrecognized or uncorrected, and diminish in significance. Mistakes, however, can occur due to confusion, uncertainty, conflict, or lack of knowledge, and the effects of mistakes may be universal and persisting. Thus, although latent mistakes and recovery slips are possibilities, they should not be as likely as latent slips or recovery mistakes. The limited evidence from real events seems to support this thesis.

A Suggested Combined Classification Scheme

Summarizing the perspectives above results in a human failure event taxonomy that reflects all of these considerations. This classification scheme is depicted in the "decision" tree of Figure 9-1, which leads to twelve categories. This taxonomy integrates easily into a PRA and is consistent with others, such as that of Embrey (Chapter 2).

Mistakes are not usually modeled in normal task activities (category 3). The driving assumption is that the plan is specified by plant management and directed by technical specifications and the decision elements of the task are, thus, highly reliable. (The 1986 accident which occurred at the RMBK-1000 reactor in

Chernobyl in the Soviet Union indicates that there may be circumstances under which this assumption might need to be reevaluated.) Unspecifiable tasks obviously cannot be modeled, although some plants may have enough data to estimate failure rates for this category (category 4). Commission failures that are due to mistakes are often not modeled (categories 8 and 12). This is because it is assumed that other failure modes dominate their occurrence probabilities and the presence of extensive plant instrumentation make their effects detectable and give rise to the possibility for mitigating the consequences of these failures.

The result is that, typically, only eight of the twelve categories of human failures need to be identified in an HRA for a nuclear power plant PRA. Another benefit from this taxonomic approach is that it not only guides the event identification effort but points to the (currently accepted) technique to quantify an event, i.e., THERP or some TRC approach.

Guidelines to System Analysts

Human failure events are identified, characterized, and quantified according to the classes in Figure 9-1. Since the effects of slips during plan-directed activities do not usually propagate unassisted by other failures beyond the component(s) affected, they can be be incorporated into fault tree models. Mistakes, which are most significant in event-driven activities, are best incorporated at the top of or above the system fault tree level because of their propagative potential. Recovery events are tacked on the end of the dominant sequences for which they apply and may not be formally included in the event or fault trees. In this way, the classification effort supports the integration effort.

As has been mentioned, the human reliability analyst, while an integral part of the PRA team from the beginning, plays an initial supporting role to the systems analysts. This is because it is the systems analyst who has the initial responsibility for the development of the risk models. This being the case, in order to ensure that human failures are given close consideration, the human reliability analyst must provide guidance to the systems analyst in the fault tree and event tree development. The guidance given can take several forms, from casual contact to continual interface and review. However, experience has indicated that the best balance is provided by selected human reliability training, combined with the application of formal guidelines, and provision for regularly scheduled periodic reviews of the developing models.

Consistent with this approach, the following guidelines are provided to the developer of fault trees, special initiator models, and event tree logic to properly identify human failure events. The guidelines are a simplification of the taxonomy

Qualitative Analysis 91

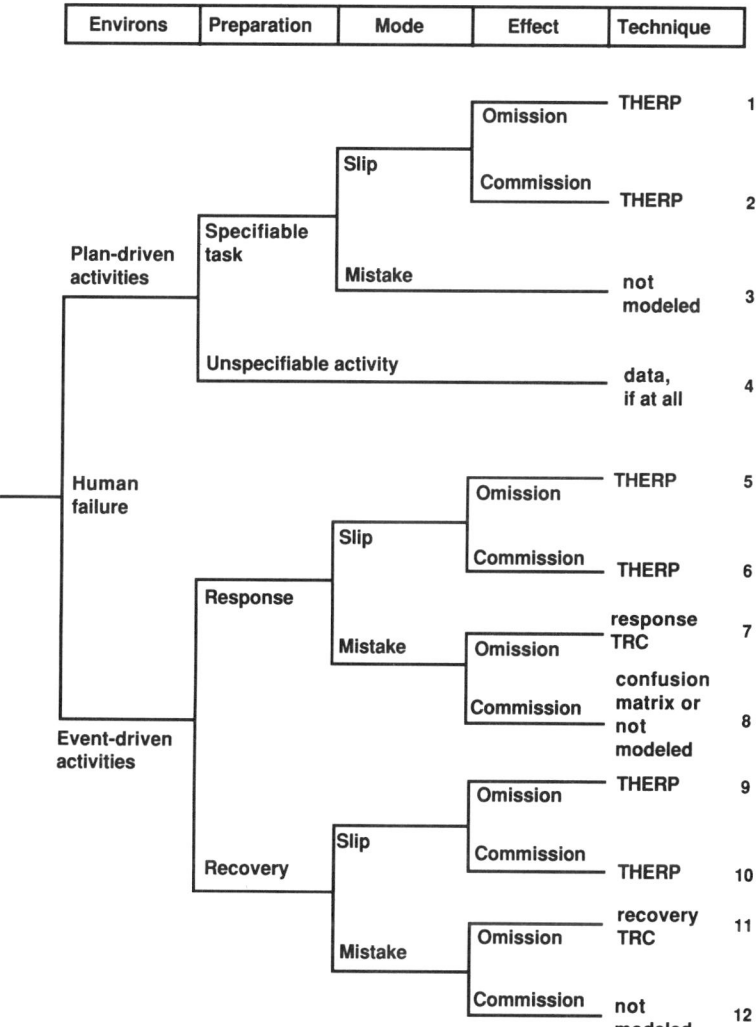

Figure 9-1. Classification system for human failure events.

developed in the previous section. Proper identification in this case means:

1. Identifying those failure events appropriate to the sequence(s) in which the events will be factored and

2. Locating those failure events in the appropriate place in the logic tree system.

Although the "human system" is radically different from hardware systems, the logic structures should not be cluttered with imaginative but unsupportable human

92 Human Reliability Analysis

failures. These guidelines are intended to assure proper human failure event identification.

Figure 9-2 is a translation of Figure 7-2 into terms related to the analysis elements of a (level I) PRA. Human actuation of equipment and test and maintenance activities can result in latent failures of incident or accident management systems or directly lead to initiators and challenges to the incident management systems. The former kind of event is best modeled in the fault trees and the latter are usually subsumed in the data analysis needed to quantify initiators. Human actuation or recovery activities can result in post-initiator, i.e., incident and accident management activity, failures. These events are usually best modeled at the event tree levels, i.e., the top logic or the tops of the system fault trees, or, in the case of recovery events, left as adjoints to cutsets.

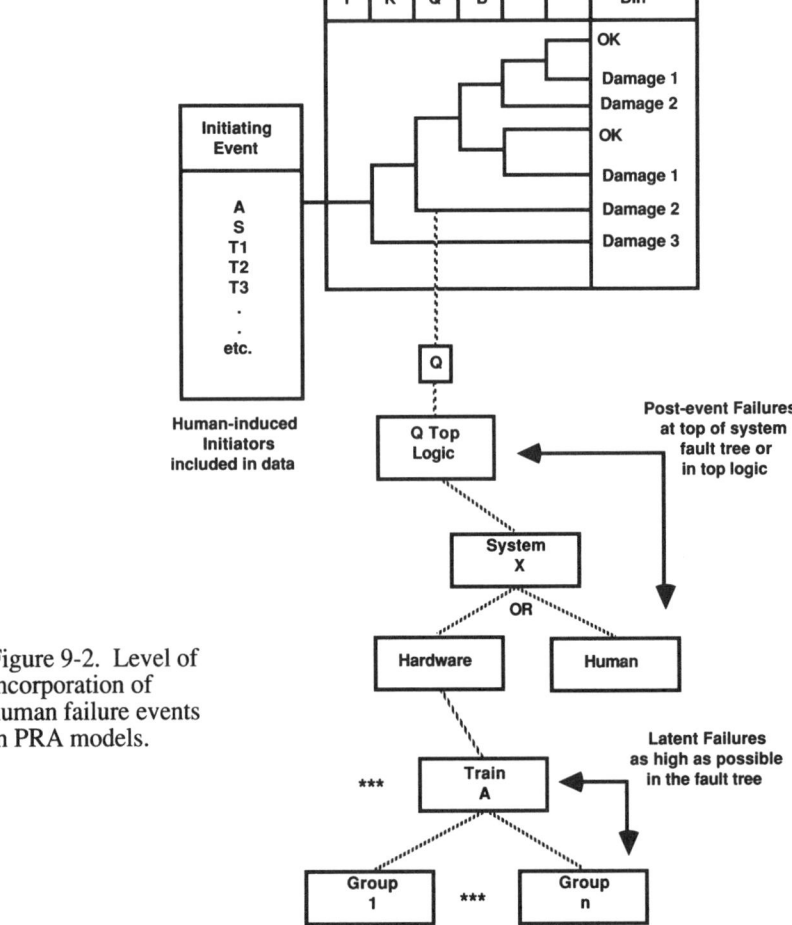

Figure 9-2. Level of incorporation of human failure events in PRA models.

The reason for leaving the recovery events out of the (initial) plant models is twofold. First, a recovery event is highly dependent on information that can only be gleaned from the specific details of the sequence cutset and phenomena in which the recovery is postulated. For example, the ac failure of a motor-driven pump may in some sequences mean a recovery event of the operators' trying to switch to another ac power bus and restarting the pump. But, if the sequence is a station blackout sequence, i.e., all ac is lost, then recovery is quite different. The fault tree analyst cannot anticipate such sequence specifics and should not, therefore, model recovery events.

The second reason for leaving recovery events out of the models is economic. A recovery event necessarily introduces an "and" gate logic, i.e., something fails *and* the operators fail at a recovery attempt. "And" gates multiply the number of cutsets enormously, until no current computer can solve the sequence models. Leaving out recovery events, which can then be reintroduced after the cutsets are solved assures a PRA solution and reduces its cost.

Guideline 1. The systems analysts should describe human failure events functionally only. For example, "crew fails to switch to ECCS recirculation" is sufficient, where "crew fails to open valve ECCS-X" is too specific relative to equipment and human mechanism.

Guideline 2. Only three types of human failure events should be identified— failures before an initiator, called latents; human-induced initiators; and post-initiator human failures.

Guideline 3. All latent human failure events should be incorporated in the appropriate fault tree at the highest level that accounts for all of the equipment manipulated during the test or maintenance task. The highest level is typically the train or part-train level (as indicated in Figure 9-2). A task is the sum of all the actions that lead to a test or maintenance objective, including system isolation, the intended activity, and system restoration. An event should not be decomposed into individual failure events for each piece of equipment manipulated in the task. Although such a practice is a structured and natural way to identify human failure "events", it can result in meaningless cutsets at the system and sequence levels.

Guideline 4. Human failures that lead to an initiator should be assumed to be included in the data for the initiator class that includes the human failure and not modeled explicitly. In the case that a special hardware initiator is modeled, the fault tree can include latent error events identified according to Guideline 2. However, no human failure event should be identified that will require quantification with a frequency, i.e., a probability per unit of time, rather than a probability, unless data is able support such a quantification. For example, an event like "operators cause trip while doing so-and-so" cannot be quantified other than by using data, which will be typically insufficient to produce such an estimate.

Guideline 5. All post-initiator human failure events that relate to the manual actuation of a system should be modeled as a single event for that system in an "or" gate at the top of the system model. The project may alternatively model such an event in the top logic of the appropriate event tree(s) (see Figure 9-2). The functional description of this kind of event is "crew fails to initiate X".

Failure events of the type: "crew inadvertently manipulates equipment X", should not be modeled. These events have not been shown to be general contributions to risk in the nuclear power plant operation setting. If a special situation that could induce a human failure of this type is found in the analysis of a system, this information should be relayed to the HRA team but not modeled.

A type of post-initiator human failure event is sometimes referred to as a commission error. This type of failure can be generically described as: "operators operate system X too early" or "operators operate system Y rather than X". Such events should not be routinely located in the logic structure. Again, if a special situation is noted that could give credence to such an event, this information should be relayed to the HRA team.

Guideline 6. Where a human action is not required for system actuation, for example, manual initiation of scram, a corresponding human failure event is logically associated in an "and" gate with the failure event of the system's actuation. These events are often termed recovery events and amount to failing to restore the function of faulted equipment or failure to find alternatives to its function. Such recoveries may be identified and described early in the systems analysis but should not be modeled until the total sequence cutset has been identified. At this point a recovery analysis will identify the appropriate recovery actions, given the sequence context. A proper functional description of a recovery failure is "crew fails to recover from sequence ABC".

Guideline 7. The HRA needs supplemental information from the system analysts to quantify any human failure events identified by the systems analysts. This includes at a minimum the information in Table 9-1.

Example

To provide some insight into the process of identifying and incorporating human failure events into a PRA, the following example is provided. The situation is that the plant is a PWR that has suffered a loss of coolant accident (LOCA). Initially, emergency core cooling system (ECCS) water (high and/or low pressure injection) have successfully performed their function. Some 30 to 100 min into the sequence, the tank source of borated injection water will deplete. Typically, a PWR has the

Table 9-1
HRA Information Needed from System Analysts

Event type	Needed information
Latent Failures	
Miscalibration	Calibration procedure number Equipment affected
T&M restoration failure	Procedure number(s) Equipment affected
Post-initiator Failures	
Manual initiation failure	System or portion of system affected Actuation or operation procedure number
Recovery failure	Equipment to be restored or realigned Location of equipment to be manipulated

option of switching to sump suction to obtain the water that has spilled from the system because of the LOCA and to recirculate it back through the core. This is a closed system relative to the containment building (although, of course, not relative to the reactor vessel) and has two missions. The water must be cooled by means of heat exchangers to provide sufficient heat transfer temperature difference and the flow of water must be sufficient to cover the core and prevent damage. Typically, two trains of pumps and associated valves are designed to accomplish ECCS injection, with the flow equivalent to one train necessary for successful recirculation.

The activity to accomplish recirculation involves opening the normally closed valves that connect the recirculation pumps to the sump, actuating the low pressure pumps (if they are not already on) and assuring their operation, and then closing valves to isolate the recirculation system from the emptying tank. The tank level is monitored by instrumentation and sounds an alarm at a low level and a low-low level. Procedures typically call for the crew to begin alignment and assure the operation of the low pressure pumps on the low level alarm and to have the system functioning properly by the low-low level alarm. There is at least a 15 min duration from the low-low level alarm until the tank empties under all possible variations of

96 Human Reliability Analysis

the LOCA. These actions are highly emphasized in training and are simulated in the plant simulator (if there is one).

Failure or unavailability of ECCS recirculation, U_R, is a direct path to core damage, even with successful initial cooling. Thus, one possible PWR core damage sequence is

$$T_L U_R$$

where

T_L represents the LOCA
U_R represents the failure of recirculation

The juxtaposition of the two events means that both must occur for this sequence to occur and the resulting core damage to occur. This representation is only functionally described and must next be described in more detail. Figure 9–3 is a fault tree model of the event U_R. The initiator is usually not modeled but is quantified using an analysis of industry and plant data. The fault tree is in the initial top logic or is a system model, i.e., those events that are "pre-recovery". The cutsets, i.e., failure sets, that are logically obtained from this model are listed in Table 9-2. Each of these representations is a possible variant of the basic sequence representation, with more detail of its description. [Note that this tree is not accurate and is only provided for illustration purposes.]

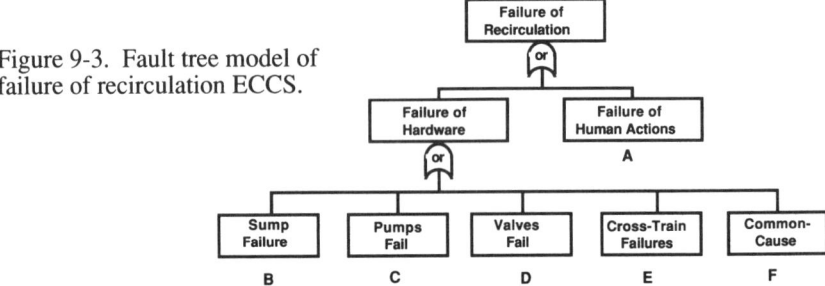

Figure 9-3. Fault tree model of failure of recirculation ECCS.

The first cutset includes the human failure event, A, functionally described as the "crew failure to achieve recirculation". This event is modeled because the actuation of the low pressure pumps and the alignment of the system away from the tank to the sump are designed to be manual actions. The other cutsets are purely hardware cutsets or may include fault modes that are latent human failures, e.g., a sump valve is left unavailable due to an error in maintenance. Relative to the operators facing the sequence in real time, whether equipment is failed because of hardware faults on demand or human faults in the past makes no difference.

Table 9-2
Original Cutsets for Recirculation Failure

Cutset	Recovery Potential
T_LA	none - the recovery is assumed to be integrally part of the human failure event, A
T_LB	containment spray system pumps, the sump, and crossconnect valves can be used to refill the tank
T_LC	spray pumps may serve core cooling as well with realignments
T_LD	manual actuation may be possible
T_LE	crossconnect valves may be opened
T_LF	may be none

The specific fault mechanism, however, influences the possibility of restoration of the equipment or recovery by using alternatives that will produce equivalently cooled flowing water from the sump source. Table 9-2 also shows the recovery possibility of each cutset. Each recovery opportunity may identify one (or more) human failure event that is formally part of the fault tree in Figure 9-3 but which may be only documented with the final sequence cutset descriptions. One of the events in the fault tree is examined more closely next.

Event E represents mixed-train failures of recirculation, e.g., one outlet valve in one train and the pump in the other train fail. Plants typically have crossconnect valves that allow water to enter one train and exit the other for test, maintenance, or startup functions The crossconnect valves are normally closed but can be opened (often with the valve handwheel). An alternative to a single train of ECCS, thus, is sufficient parts of two trains. Operator training in most plants specifically includes these contingency actions; procedures may or may not explicitly include this activity.

First, event E is adjoined with E_R, the recovery failure model. [Again, this may not be done formally.] The recovery model includes the possibility for hardware failures, E_{Rhdw}, and for failures in the human recovery activities, E_{Rhum}. Since either type of failure fails recovery, they are related in an "or" gate. These new events extend the model in Figure 9-3 to that in Figure 9-4. Emulating Table 3-2, Table 9-3 identifies the likely human failure modes, mechanisms, and causes for E_{Rhum}. So far, the classification tree of Figure 9-1 has been traced through the event-driven and recovery branches and now both the slip and mistake modes are seen to be possible

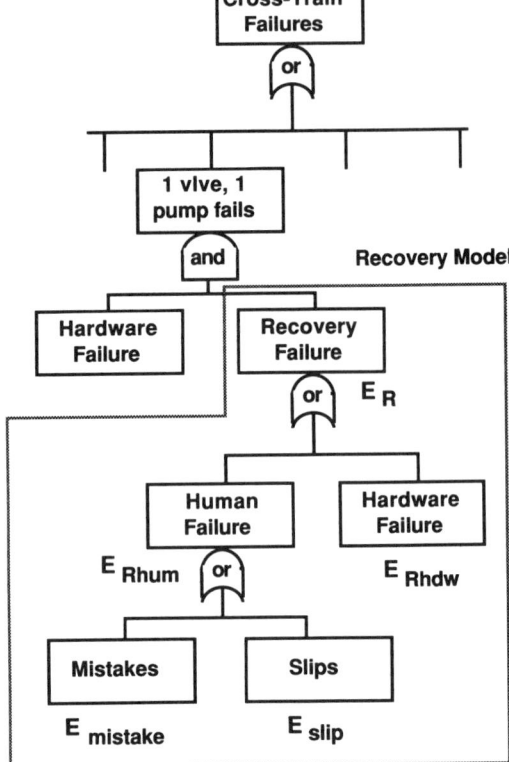

Figure 9-4. Recovery fault tree model for recirculation failure event E.

failure modes. These two modes are designated E_{slip} and $E_{mistake}$. Finally, there seems to be no realistic commission potential for the mistake mode but both omission and commission potential for the slip mode. The final events are designated as E_{slipO}, E_{slipC}, and $E_{mistake}$, which completes the human part of the recovery model. Obviously, the hardware failures of the recovery model induce the possibility of further recovery activities, but the cutset frequencies of these subsequent perturbations would be dominated by the frequencies of the modeled events and their associated cutsets. (This is ultimately a decision of the PRA manager, not the human reliability analysts).

In this manner, all human failure events are identified and incorporated into a PRA. There are shortcuts that come with experience, but the conceptual methodology stays intact.

Identifying Reliability Influences

Once a human failure event is identified, classified, and incorporated into the PRA, the final effort of the qualitative analysis is to identify the likely major influences

Table 9-3
Human Failure Event Decomposition

Mode	Mechanism	Cause	Effect	Designator
mistake	recognition fault	anticipation of at least 1 train's success	omission	$E_{mistake}$
	misidentification of alternative	not well-trained	omission	$E_{mistake}$
slip	locate incorrect valve	inexperience	omission or commission	E_{slipO} E_{slipC}
	manipulate valves incorrectly	bad labeling	commision	E_{slipC}

on the reliability of the event. Because human behavior is actively diverse, flexible, and adaptive, the converse also holds true: human behavior can be readily influenced, for better or often worse, by numerous environmental factors. THERP documents and classifies the many possible influences on human reliability. However, experience with several PRAs and their accompanying plant reviews indicate that few categories play any significant role in an HRA (see Table 9-4).

These influences can be first grouped according to whether they affect the behavior of an individual or arise only in group activities (Table 9-5). The primary categories of individual influences are those that effect the person because of the mechanisms and performance attributes that make up that person. Other influences come from the environment in which the person must perform. Since all groups are groups of individuals, these influences carry over. However, communication and dynamic factors can further influence the performance of several people and their interactions.

SLIM

In today's state-of-the-art in HRA, the associating of an influence with a human failure event necessarily depends on the judgment of the analyst. This judgment can suddenly "appear" in the analysis or can be "exposed" in some systematic way. Judgment is notoriously unreliable when about things never experienced, yet the incorporation of the knowledge and related experience of people associated with a nuclear power plant is essential for a credible HRA. This has led to the systematic

Table 9-4
Significant Influences in Prototypical Situations

Reactor Type	Activity or Sequence Type	Major Influence
BWR	transient without scram	time SCLCS conflict new procedures credibility
	manual ADS actuation	time competing resources conflict new procedures specific failure configuration
	long-term cooling	competing resources specific failure configuration
PWR	feed & bleed	time conflict new procedures indication of critical parameters competing resources credibility
	SGTR	time accident signature competing resources
	ECCS recirculation	time competing resources new procedures tank level indication size of LOCA
LWR	loss of all ac	time competing resources manpower allocation TDP control
	loss of all dc	time confusion/misleading indications competing resources
	major fire	time competing resources firefighting manpower remote activities

Table 9-5
Significant Influences on Human Reliability

Influences that Affect the Individual

    Organism
- behavior type
- skill level
- training adequacy
- stress reaction

    Environment
- time constraints
- burden
- procedure type and adequacy
- adequacy of I&C
- physiological factors
- job and shift factors

Influences that Affect Groups

    Communication
- remoteness of crew
- communication equipment

    Dynamic
- structure and role of crew
- dependency of crew actions

technique to incorporate judgment into HRA, originally called the success likelihood index methodology (SLIM).

SLIM (Embrey, 1983) was adapted from techniques used in the so-called decision theory sciences. Its intent was to structure the input from plant risk experts for risk-related human failure events into a relative ordering based on a ranking index from 0 to 1. SLIM attempts to integrally weigh and rank the major influences on an event and produce by indirection the SLI, since direct probability estimation is often inconsistent and inaccurate (Stillwell, 1982).

There are essentially three types of influences relevant to the way they influence an event. Some influences are always "bad" if they are present at all. Conflict is such an influence, since it can never help when it is present at all. Other influences are always good, for example the fact that an important piece of equipment has a direct, instrumented indication of its operational status that cannot be misleading is good. The technology used to indicate the status may be a red/green light system or a digital speedometer and thus may be better in some instances but always good. Finally, some influences may range from "bad" to "good". For example, a procedure may be written so as to confuse and generally inhibit successful action or may be written so as to serve as a critical mnemonic aid in an infrequently occurring situation.

Figure 9-5 indicates a way in which the SLI can reflect these three types of influences, with 0 representing the "worst" possible situation relative to successful action and 1 representing the "best" possible situation. The fact of an always bad or always good influence is represented by restricting the range from 0 to 1 to 0 to 0.5 and from 0.5 to 1, respectively.

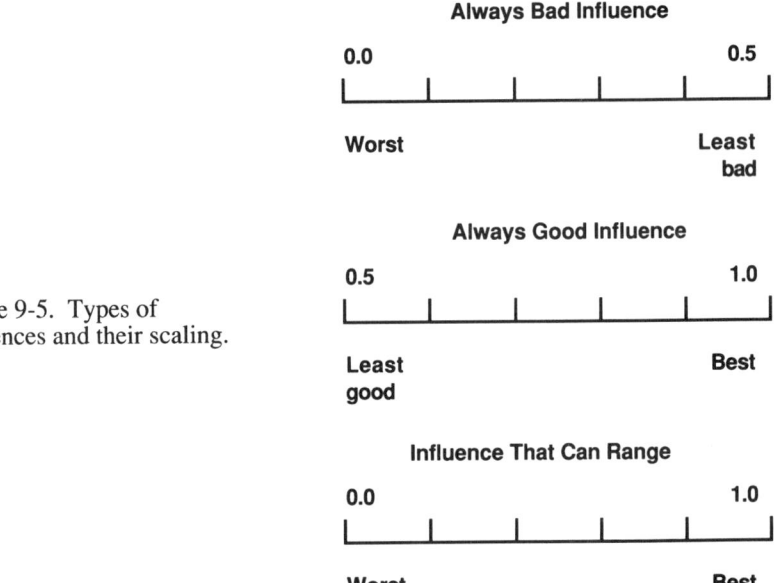

Figure 9-5. Types of influences and their scaling.

If a human failure event were to have only a single influence, then its rank on one of the scales in Figure 9-5 would be the SLI. However, most events are influenced by multiple factors. A procedure is needed to integrate their common effects into a single index.

Each influence is judged as to its relative importance among all the identified influences and then assessed a measure of the degree which it is good relative to the scales in Figure 9-5. These weights are combined in the calculus of Table 9-6. The basic SLI formulation is:

$$\text{SLI} = \Sigma_i [(r_i / \Sigma_i r_i) \times q_i] \tag{1}$$

where

Σ_i is the sum over n total influences,

r_i is the rank of the ith influence, and

q_i is the quality of the ith influence.

Table 9-6
SLI CALCULUS

---

1. Choose influences appropriate to specific event and situation.

2. Rank influences as multiples of the least important for given situation, which is set to "10".

3. Sum the rankings of all influences and normalize rankings to this sum, i.e., divide each rank by the sum.

4. Assess each influence's quality, i.e., its position in a spectrum of possibilities from best (1) to worst (0). Note that "worst" means "worst licensable" not "worst conceivable".

5. Compute the "dot product" of the ranking and quality vectors.

6. This result is the success likelihood index (SLI) which can be entered into a technique that calls for it.

The SLI for a human failure event is considered by Embrey to be logarithmically proportional to the probability. The result is that a SLIM calculation requires a success likelihood index and two "anchor" probabilities that correspond to two known SLIs. Since there is no obvious way to generate the anchors, except in those unusual cases when data is available, a different approach is used (see Chapter 10).

The example from the last section is continued. The event in question, $E_{mistake}$, is a mistake in diagnosing the mixed failure and finding an alternative, the crossconnect valves. The likely error mechanisms are a potential recognition fault or the misidentification of the alternative. The major influences on this particular recirculation failure include those from Table 9-4, the generic recirculation failure influences: time, the possibility of competing resources, i.e., the ongoing workload, the effects of training on the new emergency procedures (ERGs), the indication of the tank level, and the size of the LOCA, which affects whether the low pressure pumps are already on. Time and the rule-based nature of the ERGs are explicitly factored into the system described in Chapter 10 and are not included in the SLI process for this event. The specifics of the event also suggests that the expectation of this failure is an influence and the contingency training for such events at the plant is an influence. Thus, a matrix of information for these influences is depicted as in Table 9-7.

The first influence in the table is of the always bad type; the second is of the always good type, while the last three may range bad to good. Thus, the quality of

Table 9-7
SLIM Calculation for Recirculation Event, $E_{mistake}$

Influence	Type	Rank	Relative Rank	Quality	Product	%
competing resources	bad	10	0.05	0.4	0.02	3
tank level indication	good	50	0.23	0.9	0.21	31
size of LOCA	both	10	0.05	0.2	0.01	1
expectation of failure	both	50	0.23	0.3	0.07	10
training on contigencies	both	100	0.45	0.8	0.36	54
		220		SLI	0.67	

the first influence must be no more than 0.5 and the quality of the second influence must be at least as great as 0.5. The other three quality factors are unrestricted in the interval from 0 to 1.

The influences are ranked first, starting with the least important one(s) to the event, which is ranked a value of 10. Since the recirculation failure occurs some 30 to 100 min into the accident, the workload should have diminished and not be much of an influence. Also, the need for low pressure pump operation is known so well that the size of the LOCA should make no difference (this particular version of the example will assume that the pumps are not on). So both influences are judged least important and ranked as 10. Tank level is the key objective indication of the need for recirculation and effectually actuates the recirculation part of the LOCA procedure. This is considered to be 5 times as influential as the least influential influences and is ranked with 50. The expectation of the mixed failures in recirculation equipment is fairly low and thus is judged as influential as the level indication. Finally, training on contingencies is crucial, even if the particular step is accommodated explicitly in the procedure. This influence is judged to twice as influential as the indicator and the expectation influences and ranked as 100.

These ranks are next normalized relative to their sum, 220 and column four of the table results.

Next the quality factors are induced from the judges with the restrictions mentioned. The results will look something like column five of the table. The product of the relative rank and the quality is recorded in column six. The SLI is calculated as in (1), which is the sum of column six.

Column seven calculates the relative percent contribution of each influence. Training has resulted as most influential with the tank level indicator second.

Together these two influences account for 85% of the SLI. Note that although expectation is an important influence, it does not contribute much to the SLI because of its relatively poor quality factor.

Such a SLIM system can be massaged mathematically to indicate the sensitivity of each factor input by the judges. In this way, modifications due to changes in a judge's opinion or to actual changes to the plant can equally be assessed.

The next chapter develops the approach to handling the other major influences on human reliability.

Chapter 10
A TRC SYSTEM

The ORNL simulator data presented in Chapter 5 became available just when the applicability of the THERP methodology (as it existed until about 1983) was being questioned, particularly as it related to post-initiator human failures in events such as occurred at TMI. The common concern was that the operators at TMI [and others] did not understand the failures at their plant nor their resulting phenomena well enough to use existing procedures. It was generally conceded that had these procedures been used, the operators could have corrected the problems without the loss of the plant. The question raised by this event was whether some risk-significant behavior occurs "outside" a procedure system, and, if so, whether a procedure-oriented HRA method could be considered robust enough for PRA.

The solution, adopted by the founder of the THERP methodology as well as most HRA practitioners, was to "enhance" the task-analysis-oriented HRA method so that it could accommodate diagnosis and decision making errors, i.e., mistakes, as well as slips, which were known to predominate in highly-proceduralized or practiced activities. The quantitative response to this concern motivated the development of a time reliability correlation (TRC), which will be systematically elaborated in this chapter. In addition to this analytical response, the industry responded to this problem in a practical way by implementing a new philosophy to emergency procedures and thereby developing symptom-based procedures.

Basic Premise of a TRC Approach

The basic premise of a TRC approach lies in the nature of an offnormal event and the requirements it places upon the operators that must respond to the event. The occurrence of an offnormal event:

1. forces the operators of a plant to respond to conditions not of their making or intention

2. forces the operators to diagnose the situation at hand, interpret its implications on future plant operation, and to decide on a plan to respond—all in the real-time of the event phenomena

3. forces a time schedule on the operators that is uncertain in its details and can only be inferred from the pace of the changing of critical plant parameters or anticipated by analysis

4. forces the operators to succeed in their actions, since failure risks loss of plant equipment, property, or even lives.

108   Human Reliability Analysis

This event-driven regime of behavior sets in motion cognitive and perceptual and motor processes that take time to implement. Variability in specific event characteristics, individual personal characteristics, and job-related characteristics place a fundamental uncertainty on the prediction of a response. It is this source of uncertainty that is observed in simulators as distributed response times and *must be quantified with some time-dependent distribution.*

Some TRCs Used in PRA

A time reliability correlation (TRC) is but one of several psychometric techniques that have been used in this century to define and differentiate human performance mechanisms (Card, et al., 1983). PRAs since WASH-1400 (USNRC, 1975) have derived TRCs based mostly on speculation. Figure 10-1 shows the most often quoted curves. These curves are discussed in the next sections.

Each curve in Figure 10-1 is also assigned lognormal parameters of median response time and error factor in order to compare them to the simulator TRCs. These parameters can be calculated from knowledge of two points on a curve as demonstrated in a subsequent section.

The THERP Nominal Diagnosis Model

An intermediate draft of the THERP documentation provided a TRC for diagnosing a large loss of coolant accident (LOCA), which is the design-basis for an LWR. This LOCA curve (see Figure 10-1) was based solely on the judgment that the reliability of a lone operator as measured by the probability of error would be no more than

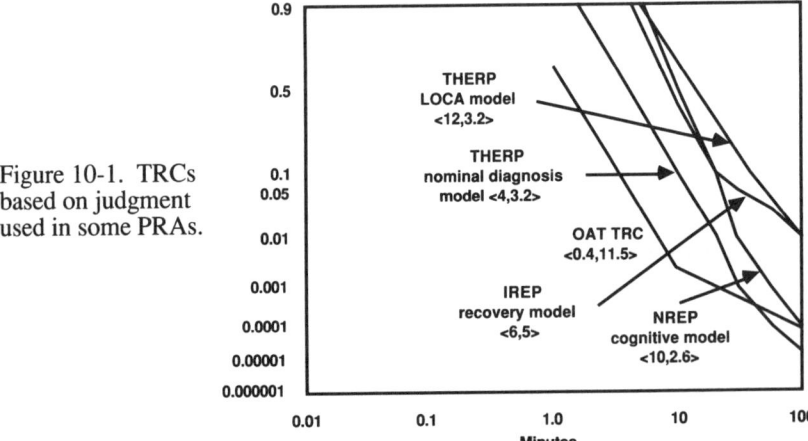

Figure 10-1. TRCs based on judgment used in some PRAs.

90% for 30 minutes into the accident and then would increase slowly afterward. If so-called threat stress was induced by the failure of automatic systems, then the reliability was assumed to reach a limit of 75% and not increase further. This latter feature first raised the issue of whether human reliability for time-dependent failure modes always increases over time or terminates or approaches some asymptotic value. As can be inferred from Chapter 4, such an asymptotic behavior is mathematically incompatible with a TRC as a CCDF. However, the hazard function is asymptotic for some values of σ (or error factor). This may mean that the intuitive measure of probability versus time is indirect, i.e., by means of the hazard function, rather than directly associated with the CCDF.

The final version of THERP added a manning model to the LOCA curve idea. The manning model is an assumption made on the number of crew members and their job responsibilities over time in a typical offnormal situation. The dependency model of THERP (see Chapter 6) is then used to develop a joint, conditional probability of the "extra" operators' failing to backup the failure of the first operator. The values of the joint probability over time is a TRC and was the basis for the "nominal diagnosis model" (see Table 10-1). This TRC was to be used to estimate the probability of errors made in deciding which procedure to use, given a situation and the diagnosis of the situation.

Table 10-1
THERP Manning Model
(From Swain and Guttmann, 1983)

Time (min)	Conditional Probability for Crew	Joint Prob	TRC Value
10	0.1 operator 1 (basic probability) 1.0 operator 2 (complete dependency) 0.55 shift supervisor (high dependency) 1.0 no credit for shift technical advisor	0.055	0.1
20	0.1 operator 1 (basic probability) 0.55 operator 2 (high dependency) 0.23 shift supervisor (moderate dependency) 0.55 shift technical advisor (high dependency)	0.007	0.01
30	0.1 operator 1 (basic probability) 0.55 operator 2 (high dependency) 0.15 shift supervisor (low dependency) 0.15 shift technical advisor (low dependency)	0.0012	0.001

In the latest variant on THERP (Swain, 1987), a mechanism is provided to "credit" the presence of the new symptom-based emergency procedures (EPs) and their effects. If a failure event involves activity that is explicitly included in the EPs, the lower bound probability for a given available time may be chosen as an estimate of the central tendency in place of the median value. This turns out to mean that the new EPs improve the reliability of some activities by a factor of 3 to 10, depending on the available time. The THERP nominal diagnosis model still has two major limitations:

1. Decision making and its accompanying burden are not modeled explicitly but are instead assumed to be covered by the uncertainty bounds on the TRC.

2. No other performance shaping factors or influences are modeled by the THERP TRC system.

Fitting the THERP TRC "data" to a lognormal form that preserves two points shows this TRC to be equivalent to a <4, 3.2> lognormal TRC. (See subsequent section on the procedure to "lognormalize" a TRC.)

The AIPA System

A PRA of the high-temperature gas-cooled reactor concept in the late 1970s included a TRC system in its basic documentation, called the Accident Initiator Progression Analysis (AIPA). This TRC system took an exponential distribution family as its basis (Fleming, et al., 1979). Thus, the only free parameter of any TRC was the mean response time. A consensus of judgment was used to estimate the mean response times for several actions. This TRC system is a precursor to the Human Cognitive Reliability model but was not used directly anywhere else. No curves are replicated from this system.

The OAT TRC

The Susquehanna PRA used a log-linear curve developed by Wreathall (see Hall, Fragola, and Wreathall, 1982) as part of the Operator Action Tree (OAT) method. This curve was developed by consensus of operations, reliability, and human factors experts and matched an interpretation of the numerical implications of TMI published along with the description of OAT. An error probability of 0.015 was assigned to 30 min and 0.0001 to 100 min based on rough estimates of industry-wide experience. Wreathall also raised the issue of whether there is a limit to time-dependent human reliability, terminating the TRC at 0.0001 or 0.00001, depending on the frequency of the initiator to which the response was made.

The OAT TRC cannot integrally address any other influences. Wreathall did recommend adjusting a TRC value by a factor of 3 or 1/3 to indicate less or more favorable conditions than "normal". Ninety-percent uncertainty bounds on the TRC values were assumed to be a factor of 10 above and below the value.

The OAT TRC is included in Figure 10-1 and is equivalent to a <0.4, 11.5> lognormal TRC by preserving the two data points mentioned above.

The NREP TRC

The NRC proposed a National Reliability Evaluation Program (NREP) in the late 1970s that would have required a level 1 PRA (see the Appendix) for all nuclear power plants. To assist the utilities in performing this PRA, Brookhaven National Laboratory developed, among other methods, a TRC that was to apply to errors in so-called cognitive behavior. THERP, as it existed then, was to be used for procedure-oriented failures.

This TRC was based on speculations from risk and human reliability experts, including Fragola and Swain, and is reproduced in Figure 10-1. The lognormalization of this TRC results in a <10, 2.6> curve.

The IREP TRC

The Interim Reliability Evaluation Program (IREP) was a four-plant program that was to serve as a precursor to NREP. Only late in its implementation did IREP attempt to handle what it termed "recovery" errors. This, too, was done by means of TRCs, one for actions that could be performed in the control room and the other for actions that had to be performed out of the control room local to equipment. The latter curve is the former curve with the time axis slipped by ten minutes to account for the extra-control room effort to locate the correct equipment.

Only the in-control room TRC is duplicated in Figure 10-1 and its lognormal equivalent is <6, 5>.

The Human Cognitive Reliability Model

The HCR model (Hannaman, et al., 1984) is a system of three basic types of TRCs, one for each of skill-, rule-, and knowledge-based behavior (Hollnagel, Pedersen, and Rasmussen, 1981). Each TRC is a representative of the Weibull family of probability distributions. The development of this model has been funded by the Electric Power Research Institute, which has jointly with the NRC and Sandia National Laboratory used the LaSalle simulator to try to "verify and confirm" the model. No results are yet available from the EPRI effort, but the Sandia analyses did not indicate any need to use a Weibull distribution family nor any justification for the three types of behavior to produce characteristic curves. [The Sandia researchers found that the fit to lognormal distributions is reasonably good (Whitehead, et al., 1987).]

The HCR system does have one feature that other previous, publicly available TRCs do not. Not only can different types of behavior lead to different TRCs, which is a general lesson gleaned from Fragola's work (Hall, Fragola, and Wreathall, 1982), but other "minor" influences can be identified and their effects quantified. These adjustments to a basic TRC accommodate one of the major criticisms of the TRC concept, namely, that a TRC models human reliability solely as a function of time. The HCR technique can be extended or an alternative can be postulated (as is done later in this chapter) to allow *any* influence on human reliability to indeed have influence.

Because the HCR model is still under development, its TRCs are not included, but may be found in the forthcoming EPRI documentation.

Except for the HCR, the above TRC systems are based on speculation rather than data. Except for its Weibull representation, the HCR model is subsumed in the TRC system that will be described in the following sections. This system is based on interpretations of simulator data and has a lognormal distribution basis. The next section discusses the lognormal family and the following section provides an interpretation of the simulator data.

The Lognormal Family

The data from simulator exercises suggest that the lognormal family of distributions is sufficient in modeling TRCs. This section describes some features of this family.

Logprobability Paper

Graph paper can be constructed that will display the distribution function of a lognormal density as a straight line. This paper is called logprobability paper because it includes as one axis the logarithm of the independent variable, which is usually time, and the dependent axis has probability as its variable, which is arranged so as to be linearly proportional to the normal (Gaussian) variates (see Chapter 4).

A TRC System 113

Response time data is plotted as described in Chapters 4 and 5. The ORNL and LaSalle data are plotted on logprobability paper in Figures 5-4 and 5-5.

Such paper gives a ready means to the analysts to discern the distribution type, estimate the median and error factor of a TRC, and to compare TRCs.

Basic Parameters and Formulation

A lognormally based TRC is completely determined by two factors, e.g., μ and σ in Chapter 4 or, more descriptively, the median response time, m, and the TRC error factor, f. The basic formulation relating the two latter parameters for a lognormally distributed random variable is:

$$t_p = mf\hat{\ }\hat{\ } (z_{1-p}/1.645) \tag{1}$$

where

$\hat{\ }\hat{\ }$	is the exponentiation operation
m	is the median response time
f	is the error factor, i.e., $f = t_{0.95}/t_{0.5}$
z_{1-p}	is the normal variate for probability 1-p, and
1.645	is the normal variate for probability 0.95.

In this formulation, a lognormal distribution is often denoted as an <m, f> curve. Note that this formulation easily arrives at the standard facts for lognormal distributions:

$$t_{0.5} = mf\hat{\ }\hat{\ }(0.0/1.645) = mf\hat{\ }\hat{\ }0 = m \tag{2}$$
$$t_{0.95} = mf\hat{\ }\hat{\ }(1.645/1.645) = mf = (t_{0.5})f.$$

The normal variates may be found in any standard statistical tables handbook (for example, Pearson and Hartley, 1954).

Lognormalizing Other TRCs

A TRC, particularly one developed from judgment, may not have an exact lognormal (or any other) distribution form. However, using (1), the TRC may be approximated by a lognormal CCDF by using two points from the TRC to solve (1) for its two unknown parameters, m and f. Table 10-2 demonstrates this process using the nominal diagnosis TRC from THERP. The net result is that the TRC is a <4, 3.2> TRC, approximately.

Table 10-2
Lognormalizing the THERP Diagnosis TRC

Available Time, t (min)	Probability of Human Failure THERP	<4, 3.2> Lognormal
5	0.9	0.4
10	0.1	0.1
20	0.01	0.01
30	0.001	0.002
60	0.0001	0.00006

Anchor 10 min value and adapt the formulation in Table 4-3 to a TRC:
$$10 = \exp[\mu + \sigma\phi^{-1}(0.9)] = \exp[\mu + 1.282\sigma]. \tag{1}$$

Anchor 60 min value and use formulation:
$$60 = \exp[\mu + \sigma\phi^{-1}(0.9999)] = \exp[\mu + 3.719\sigma]. \tag{2}$$

Divide (2) by (1):
$$6 = \exp[(3.719-1.282)\sigma] = \exp[2.437\sigma], \text{ or}$$
$$\ln(6) = 1.792 = 2.437\sigma, \text{ or}$$
$$\sigma = 0.735. \tag{3}$$

Replace σ in (2) with its value calculated in (3) and solve for μ :
$$60 = \exp[\mu + (3.719)(0.735)] = \exp[\mu + 2.734], \text{ or}$$
$$\exp(\mu) = 60 / 15.394 = 3.9.$$
Since $\exp(\mu)$ is m, m is about 4 min. (4)
Using this median of 4 is equivalent to $\mu = \ln(4) = 1.386$. (5)

Substituting μ so found into (1) yields:
$$10 = \exp[\mu + 1.282\sigma] = \exp[1.386 + 1.282\sigma] = 4\exp(1.282\sigma), \text{ or}$$
$$\sigma = \ln(2.5) / 1.282 = 0.715.$$

And using the formulation for error factor, f:
$$f = \exp(1.645\sigma) = \exp[(1.645)(0.715)] = \exp(1.176) = 3.2. \tag{6}$$

Thus, the lognormal TRC that uses the 10 and 60 min values of the nominal diagnosis model is approximately a <4, 3.2> curve. Using formulations from Table 4-2 yields the values in the rightmost column above.

Other curve fitting techniques can be used to lognormalize a TRC but this method is the easiest and preserves the fit to at least two points, which may not be the case in, e.g., a regression fit.

Adjusting a Basic TRC

The data from the simulator exercises seem to indicate one TRC is not sufficient to accommodate all event and action types. A lognormal family of TRCs can be generated with a method to "adjust" the TRC parameters based on differing conditions. One such adjustment is the asymptote of the LOCA curve and the OAT curve. As noted before, asymptotic behavior is not allowed by a CCDF. However, the slip probability in any response is non-zero and can often be used as a threshold value that stops the overall event probability from decreasing below some acceptable lower limit.

Another "adjustment" comes from the fact the TRC is used to generate a point estimate for the given available time. This point estimate has uncertainty associated with it, as do all points on the basic TRC. In effect, there is an envelope of TRCs associated with any response, one each for each level of uncertainty. The basic TRC is interpreted as the median or mean uncertainty level for this envelope. A basic TRC with its 5th and 95th percentile TRCs in shown in Figure 10-2. Such an envelope may require unique <m, f> pairs for each confidence level or may only require unique f values for each level (this latter approach would be nonsensical for probability values above 0.5, where the curves would cross).

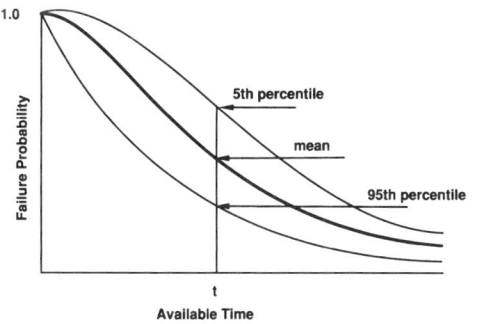

Figure 10-2. A typical TRC with two percentile TRCs to represent uncertainty.

A method to introduce other influences into a TRC system is needed. One approach is the use a success likelihood index (SLI, see Chapter 9) to interpolate between confidence bounds, say the 5th and 95th percentile TRCs. Another approach, used in the HCR model, and later in this chapter is to use a SLI-like index to adjust either the m or the f parameters, or both, to reflect the influence of influences.

The next section interpretates the simulator data that is publicly available, looking for a way to adjust the TRC system.

Interpretation of the Data

Interpretation of the simulator data can take two tactics—either a numerical analysis of the results or a classification of the results into meaningful categories and the derivation of class-average parameters. Both strategies are presented in the next sections.

Interpreting the ORNL Data

The nature of the ORNL simulator project was to discover characteristic response times for specified operator actions in the hope of establishing a basis for standards for instruments and other aids for the operator. The stochastic nature of such response times was anticipated and a probabilistic framework was established that was equivalent to a TRC. The plotting of the simulator data was made on logprobability paper (described earlier), which assumed that the resulting TRCs would be lognormally distributed. This fit was reasonably good and has the properties noted previously.

The events studied in the ORNL work are somewhat benign relative to typical plant risk. This was mostly due to limitations in the simulator. The events in Figure 5-1 are often omitted in an explicit initiator list in most PRAs of a PWR because there are much more severe but similar challenges to plant safety. For example, a steam generator tube leak, the SGTL event, is a frequent but inconsequential event compared to a full rupture of one or more tubes. Also, the small LOCA was of an equivalent diameter size that could be anticipated to be isolated easily or even ignored as slightly higher than normal leakage. However, the events in that study are precursors to more risk-significant sequences and do require operator diagnosis and actions similar to the sequences of greater risk, and, thus, do represent accident responses.

Table 10-3 shows a listing of four PWR events in ascending order of median time and a summary of a probable key contributor to the event's median response time. It appears that (not unexpectedly) the response time increases as the difficulty of the diagnosis increases from the trivial nuclear instrumentation fault to the more subtle, small LOCA.

Table 10-3
ORNL Median Times
(Extrapolated from Bott, et al., 1981)

Sequence	Critical Influences	Median Time (min)
Loss of NI	highly practiced action, core-related problem, thus high awareness, minimal diagnosis	0.5
Steam generator tube leak (SGTL)	radiation in secondary well instrumented, easy signature to diagnose, high awareness of stopping radiation leaks	1.3
Inadvertent safety injection	onset of SI obvious, must check other parameters to assure that a safety system actuation can be terminated	5.3
Small loss of coolant accident (LOCA)	slow evolving leak, can have many sources and thus, the possibility of isolation may take time	9.5

Table 10-4
ORNL Error Factors
(Extrapolated from Bott, et al., 1981)

Sequence	Critical Influences	Error Factor
Loss of NI	highly practiced action, core-related problem, thus highawareness, minimal diagnosis, no decision making	3
Steam generator tube leak (SGTL)	high awareness of stopping radiation leaks, procedures are clear on need to isolate affected SG	3
Inadvertent safety injection	check other parameters to assure that a safety system actuation can be terminated, hesitance	3
Small loss of coolant accident (LOCA)	can have many leak sources decision to isolate or go to high pressure injection	5

118   Human Reliability Analysis

Table 10-4 shows a similar listing for ascending order of the error factor. There are no difficult decisional problems involved with these sequences. Since all error factors fall in the range of 3 to 5, there is nothing to distinguish these values from the values found normally in a PRA.

The data analysis of Fragola (Hall, Fragola, and Wreathall, 1982) used the ORNL data, the data from Greene (1969), and some field data. All but the field data are reproduced in Figure 10-3. This data seems to partition the included events into three regions: one that yields median times of 2 sec or less, one between 2 sec and 1 min, and one greater than a min. These regions were once speculated to conform with the categories of skill-based, rule-based, and knowledge-based behavior. This is a basic tenet in the HCR model.

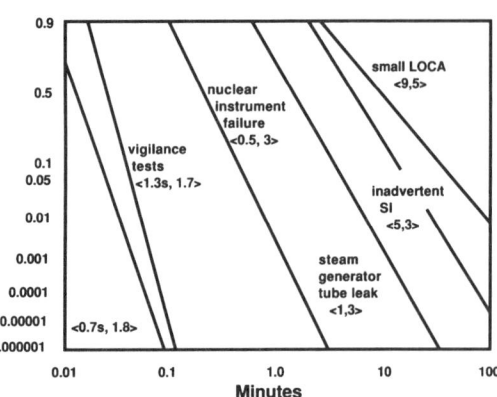

Figure 10-3. Response time data that formed initial basis for TRCs. (From Hall, et al., 1982.)

More recent research on the mechanisms of human performance do not so easily discern between rule- and knowledge-based behavior, particularly in light of the new EPs with their symptom-oriented diagnostic "rules". The middle region, however, certainly includes highly practiced activities (manual scram is also in this region according to the LaSalle data). The lower (left) region is probably the sole province of quick, motor reactions, the performance of which is the product of extensive training and high levels of skill. It is not easy to identify a task in the nuclear power plant setting that conforms to such a reaction task; military or commercial jet piloting, on the other hand, would call for a reaction TRC in this time regime.

As a result, the higher (right) region seems candidate for the whole spectrum of interesting, post-initiator operator responses (except for manual scram). The data collected to date do not eliminate the possibility that there is a rule-based versus knowledge-based region to revitalize the Hollnagel categories but, if these regions exist, they cannot be discerned from the ORNL data. (Note that the ORNL exercises were conducted well before the new EPs were implemented at TVA plants so that the new rules were not even in effect.)

Interpretation of the LaSalle Data

The event-specific TRCs from the LaSalle simulator (Figure 5-5) seem to "fall all over the place". However, when the events and their actions are classified according to generic operational characteristics and TRCs produced (Figure 5-4), the results seem generally to fall into a TRC band with median times ranging from 1 min to 20 min and error factors ranging from 3 to 5. This is quite confined from the general PRA context.

It should be noted that the aggregation of data over possibly diverse, statistical populations is a generic issue in the data analysis realm that the component data analysis tasks of a PRA must contend with as well (Fragola, 1983). In brief, typical statistical techniques tend to compress the parameters that represent the numerical uncertainty in data sets on their combination. This is because the techniques are based on the assumption that all the data sets that are being aggregated are from a larger, common population, i.e., represent the same phenomena. However, aggregation can be performed on diverse-population data sets as easily and the resulting confidence limits will shrink unjustifiably. The Sandia researchers claimed to have performed adequate statistical tests to indicate that the data could be aggregated for the action groupings chosen. If so, the "better behaved" TRCs may indeed represent actual phenomena. If not, the scattered TRCs of Figure 5-5 may lend evidence to the thesis that for every event to respond to, there is a unique TRC representing the response.

Table 10-5 lists the LaSalle general actions' TRCs in ascending median times and a summary of the probable key contributor to the event's response time.

Action 2 is an interesting datum[*]. It has been conjectured that the pace of an accident may be the driving influence in median response time. Pace can be thought of as the speed with which critical parameters change in value toward an indication of impending problems. Conceptually, pace may be equivalent to translating a faster TRC by some time lag, as:

---

[*] Action type 2 includes (solely) the activities of deciding to depressurize to low pressure and use the low pressure systems following a failure of all high pressure systems. The calculated TRC used an available time that started accruing time at reactor trip. This is clearly not when the decision needs to be made, since there will be a lag-time from the loss of high pressure water injection until the reactor water level descends to approximately the top of the active fuel (TAF) which is the EP symptom (cue) for resorting to low pressure systems. This "extra" time will likely be filled with attempts to restore some high pressure system, and the "real" TRC for action category 2 may have a shorter median time, being cued by reaching TAF.

Table 10-5
LaSalle Median Times
(Extrapolated from Whitehead, et al., 1987)

Group	Critical Influences	Median Time (min)
6	manual reactor scram trained on all scram conditions, high awareness, minimal diagnosis	0.1
4	electrical power requirements readily known	1.4
8	awareness of loss of all critical equipment in a station blackout is readily known and easily diagnosed	1.4
1	automatic system actuation conditions highly trained, operators like to feel in control	1.6
3	anticipated role as backup to automatic equipment,	2.3
7	must check other parameters to verify that automatic actuation is not necessary	3.8
9	need to verify alternatives are not better than restoration, recognize system interactions, locate room of equipment and its access	7.1
2	high awareness of need for low pressure systems as redundancy to high pressure systems, diagnosis is easy though not routine	8.9
10	must check and recheck supposedly false indicator against other parameters	10.5
5	offsite electrical equipment much less familiar and complex, diagnosis and troubleshooting needed	11.2

$f_1(t+\delta t) = f_2(t)$ for all $t \geq 0$ (3)
where
       f_1 is a TRC that is not slowed by pacing,
       f_2 is a TRC influenced by symptom pace, and
       δt is some lag time that compensates for the pace effects.

Unfortunately, this formulation does not generally hold true for probability distributions. Pace will have to be reflected in the parameters, m or f.

Other than this anomalous action type, all the actions with less than 4 min response times seem to include easily diagnosable events (see Table 10-5). The last four action types have some complicating factor up until type 5 (finding failures in the the switchyard or other offsite-source electrical power equipment),which may be in specific cases impossible to diagnose. Thus, the complexity of diagnosis may correlate fairly well with the median time to respond.

Table 10-6 lists the LaSalle general actions' TRCs in ascending error factors and a summary of the probable key contributor to the event's error factor. Again, error factors below 5 probably do not require explanation but merely represent an expected amount of human variability in uncertain conditions. This means only the last six actions need scrutiny. Each in fact seems to have varying, and maybe ascending, degrees of decisional conflict—from the concern of whether the crew should wait on an automatic actuation of a system and not anticipate it, to whether the plant is indeed down to its last recourse. This last recourse decision making in the group 8 action set was operationally similar to group 2, in that each situation

Table 10-6
LaSalle Error Factors
(Extrapolated from Whitehead, et al., 1987)

Group	Critical Influences	Error Factor
2	high awareness, focus of new emergency procedures	1.6
10	some hesitance to override indicators	2.4
4	high awareness, root cause can be questionable	3.4
6	manual reactor scram trained on all scram conditions	4.2
1	conflict: should wait for automatic actuation?	5.1
5	conflict: is this a problem with the grid that others will fix?	5.3
3	conflict: should this equipment be restored or should alternatives be found?	5.7
9	conflict: should crew be sent out of control room and actions be conducted remotely out of province of control room operators?	6.6
7	conflict: is automatic system correct and operator diagnosis not?	7.2
8	conflict: is plant really down to last resort systems, these actions will assure lengthy shutdown?	45.9

called for the depressurization of the reactor in order to use some low pressure water source. The Sandia statistical tests showed these actions were not of the same populations. The Sandia analysts conjectured that the group 8 TRC differed from 2 because of the hesitancy due to the dire conditions of a station blackout. This appears to be the only instance in the LaSalle data analysis that indicated the presence of burden. However, this examination of the data indicates that the decision making aspects of an action, especially conflict, seem likely correlates to the error factor of a TRC, or the dispersion of its basic data. Thus, what will be called burden in Chapter 12 may be the critical determinant for rule-based and knowledge-based TRCs—diagnostic burden affecting the median response time and decisional and command burden affecting the error factor.

Solution Rates

A final analytical look at the LaSalle data included the plotting of three of the group action TRCs. The groups chosen were 2, 3, and 8 (see Table 5-2); the choice was made based solely on the scatter of median response times, from 1 to 9, and the scatter of error factors, from 1.6 to 46. This is as wide as the total in a minimum number of curves that includes a "nominal" curve (group 3).

Figure 10-4 displays the solution rate function curves (see Chapter 4 for an explanation). The tight curve has an error factor of less than 2. Its peak cannot be calculated even using normal probability tables with ten-place decimal accuracy.

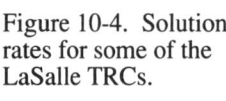

Figure 10-4. Solution rates for some of the LaSalle TRCs.

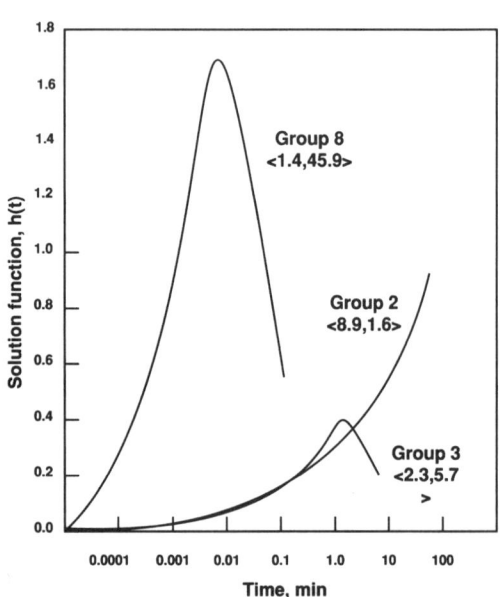

The other two curves have maxima in their solution rate curves. The error factors in these curves are 6 and 46. Figure 10-4, like Figure 5-4, shows the group 8 (last resort action) curve to be an anomaly among the LaSalle data.

Synopsis of the Patterns

An interpretation of the LaSalle and ORNL data suggest that the lognormal family of distributions are sufficient for TRCs. Thus, a TRC can be characterized completely as a <m, f> curve and fits all the requirements of Chapter 4.

It is not clear yet whether several curves are needed to reflect all the properties of all risk sequences in a nuclear power plant. The LaSalle aggregations are encouraging in that they appear to support a minimal number of curves. Burden in the form of decisional conflict and diagnostic complexity seem to be captured as variations on m and f. Decisional burden seems best associated with error factor and diagnostic burden seems best associated with the median. A way still needs to be arranged to let other influences be reflected in the TRC.

The human interaction component of the manual scram event that was a concern in the WASH-1400 PRA and subsequent ATWS regulation studies now seems a moot point. If the <0.1, 4.2> curve indeed represents an industry-applicable TRC for manual scram, then this event can be ignored. Its TRC value is less than 0.00002 after but 5 min and all scram sequences are forgiving enough to allow at least 5 min to act. Since all the other TRCs have medians of about a min or greater, the first two regimes in Figure 10-3 can be ignored in nuclear plant PRAs.

This means that the range of means is from 0.8 to about 12 and the range in error factors is from 1.3 to about 30. Using the nominal diagnosis TRC parameters, 4 and 3.2, as baselines, this range can be expressed as an exponential function of a ranking index (à la the success likelihood index) from 0, for the worst case, to 1, for the best case. That is:

$$m = 4 \exp(aw + b) \quad \text{where } w \in [0,1], \text{ and} \quad (4)$$
$$f = 3.2 \exp(cz + d) \quad \text{where } z \in [0,1].$$

Table 10-7 shows the calculations that get the coefficients a, b, c, and d. The table shows that the THERP curve has a poor median (with respect to a fast response time) and a good error factor.

The next section provides a slightly less general but similar TRC system which has been used in many PRAs to quantify time-dependent human failures.

Table 10-7
Parameterizing the TRC Range

Medians

$m = 4\, e^{aw + b}$ where $w \in [0, 1]$ and 0 is "worst"
nominal median is 4 from THERP TRC
best curve median is 0.8; worst curve median is 30, thus,

$12 = 4\, e^b \Rightarrow$ $b = \ln(12/4) = 1.10$
$0.8 = 4\, e^{a+b} \Rightarrow$ $a = \ln(0.8 e^{-1.1}/4) = -3.62$

$m = 4\, e^{1.11 - 3.62w}$ $w \in [0, 1]$

Error factors

$f = 3.2\, e^{cz + d}$ where $z \in [0, 1]$ and 0 is "worst"
nominal error factor is 3.2 from THERP TRC
best curve error factor is 1.3; worst curve error factor is 30, thus,

$30 = 3.2\, e^d \Rightarrow$ $d = \ln(30/3.2) = 2.24$
$1.3 = 3.2\, e^{c+d} \Rightarrow$ $c = \ln(1.3 e^{-2.24}/3.2) = -3.14$

$m = 3.2\, e^{2.24 - 3.14z}$ $z \in [0, 1]$

Note

for $m = 4$ for $f = 3.2$
$4 = 4\, e^{1.1 - 3.62w}$, or $3.2 = 3.2\, e^{2.24 - 3.14z}$, or
$0 = 1.1 - 3.62w$, or $0 = 2.24 - 3.14z$, or
$w = 0.30$ $z = 0.71$.

Mathematical Formulation of the TRC System

The formulation for a time reliability correlation used to quantify mistakes is a multivariate lognormal distribution (USNRC, 1983). Its formal representation is a random variable, T, that accounts for the time needed to successfully respond to the situation, i.e., without an unrecoverable mistake:

$$T = \tau_R \times \tau_U \quad (5)$$
where
- τ_R is a lognormal random variable with median of m and error factor f_R to account for the uncertainty of the process
- τ_U is a lognormal random variable with median of 1 and error factor, f_U to account for the uncertainty in the model.

The complementary cumulative distribution function (CCDF) of this random variable is the TRC. Note that the resulting distribution is no longer lognormal but can be denoted in an analogous manner as $<m, f_R, f_U>$.

The first random variable of (5) represents the response process and has four factors:

$$\tau_R = k_C k_I \tau \times 1_R \quad (6)$$

where
- τ is the median response time
- k_C is a factor to adjust τ by as much as 2 and as little as 1/2 to account for taxonomic considerations
- k_I is a factor to adjust t by as much as 2 and as little as 1/2 to account for influences or performance shaping factors
- 1_R is a lognormal random variable with median of 1 and error factor, f_R.

This random variable is distributed lognormally with a median response time of $k_C k_I \tau$ and an error factor of f_R. The adjusted median response time, $k_C k_I \tau$, is equal to m. The random variable of (6) can be inferred directly from simulator exercises or estimated as below. The correlation between time and probability in (6) can be numerically expressed as in (1).

In this way, a TRC for a special response can be developed from an estimate of its adjusted median time to respond and an estimate of the spread on the estimate.

Formulation of Mean and Uncertainty Estimates

Formulation (5) can be used to obtain a CCDF that represents the probability that a response is not successful by a given time, t. An estimate of the mean of this underlying distribution is given as:

$$E\mu = \Phi[-\ln(t/m)/\sqrt{(\sigma_R^2 + \sigma_U^2)}] \tag{7}$$

where

Φ = the standard normal (Gaussian) cumulative distribution
σ_R = $\ln(f_R)/\Phi(0.95)$
σ_U = $\ln(f_U)/\Phi(0.95)$.

The 5th percentile estimate is

$$p_{0.5} = \Phi[(-\ln(t/m) + 1.645\,\sigma_U)/\sigma_R]. \tag{8}$$

The 95th percentile estimate is

$$p_{0.95} = \Phi[(-\ln(t/m) - 1.645\,\sigma_U)/\sigma_R]. \tag{9}$$

Formulæ (7)-(9) are time-dependent and produce TRCs that estimate the mean, 5th percentile, and 95th percentile estimates for time, t. An algorithm to compute these values has been incorporated into the ORCA code (see Chapter 11).

Adjustment Factors

The formalism for computing a probability from the SLIM and TRC concepts has been presented in (6). What remains is a way to calibrate that form to specific situations.

First, the base TRC, neglecting model uncertainty, is derived from the nominal diagnosis curve (Swain and Guttmann, 1983). This curve is converted to a lognormal CCDF by preserving its 10 min and 60 min values and solving for m and f. The result is a lognormal <4, 3.2> curve, i.e., τ is 4 and f_R is 3.2 in formulation (6). This curve is assumed to apply to responses dominated by diagnosis, not aided by rules (as in the new emergency procedures), because the median is relatively high, and not dominated by burden or some other source of hesitancy, because the error factor is relatively low. Typical events of this category are recovery events not explicitly covered by procedures but taught in training. This curve is referred to as the "recovery TRC, without hesitancy".

Second, whether the event occurs while the crew is guided by a rule determines k_C:

$$k_C = 1 \quad \text{if no rule is available} \tag{10}$$
$\,0.5 \quad$ if a rule is available.

Formulation (6) reduces to the first TRC with no rule available, since k_C is assigned a value of 1. Otherwise, with a rule, the median response is (arbitrarily) halved.

Third, if hesitancy is present (due to conflict, burden, uncertainty, etc.), then f_R, the process uncertainty, is assumed to increase. To reflect this, the error factor, f_R, is doubled to 6.4 (again arbitrarily).

Fourth, a success likelihood index, x, is assumed to halve the median response time at its best (SLI=1) and double the time at its worst (SLI=0). The SLI is logarithmically factored into k_I as:

$$k_I = 2^{(1-2x)}. \quad (11)$$

These considerations fully specify formulation (6).

Finally, since there is no evidence of or apriori assumption to determine τ_U in (5) at this time, the following calibration is used. A "good" plant typically has a SLI of about 0.7. A sixty-minute, rule-based response under this SLI should result in some industry-accepted probability. The value 10^{-6} (0.000001) is chosen as this value. This results in a f_U of 1.68.

The full system described above is implemented in a software product called ORCA (see Chapter 11). Alternatively, Tables 10-8 through 10-11 can be used to interpolate mean values not in the tables. Note that 2E-3 is shorthand for 2×10^{-3}. Figure 10-5 shows the THERP TRC and two of the four TRCs developed above in the TRC range observed in the simulator data.

Table 10-8
**Time Reliability Correlation Values
Rule-based, without hesitancy <2,3.2>***

Time (min)	Success Likelihood Index				
	0.1	0.3	0.5	0.7	0.9
5	3E-1	2E-1	1E-1	6E-2	3E-2
10	9E-2	4E-2	2E-2	8E-3	3E-3
20	1E-2	5E-3	2E-3	5E-4	1E-4
30	3E-3	9E-4	3E-4	6E-5	1E-5
60	1E-4	3E-5	6E-6	1E-6	2E-7

* <m,f> stands for a lognormal distribution with median response time of m and an error factor of f.

Table 10-9
Time Reliability Correlation Values
Rule-based, with hesitancy <2,6.4>*

Time (min)	Success Likelihood Index				
	0.1	0.3	0.5	0.7	0.9
5	4E-1	3E-1	2E-1	2E-1	1E-1
10	2E-1	1E-1	9E-2	5E-2	3E-2
20	7E-2	4E-2	3E-2	1E-2	8E-3
30	3E-2	2E-2	1E-2	6E-3	3E-3
60	8E-3	4E-3	2E-3	9E-4	4E-4

Table 10-10
Time Reliability Correlation Values
Recovery, without hesitancy <4,3.2>*

Time (min)	Success Likelihood Index				
	0.1	0.3	0.5	0.7	0.9
5	7E-1	5E-1	4E-1	3E-1	2E-1
10	3E-1	2E-1	1E-1	6E-2	3E-2
20	9E-2	4E-2	2E-2	8E-3	3E-3
30	3E-2	1E-2	5E-3	2E-3	5E-4
60	3E-3	9E-4	3E-4	6E-5	1E-5

Table 10-11
Time Reliability Correlation Values
Recovery, with hesitancy <4,6.4>*

Time (min)	Success Likelihood Index				
	0.1	0.3	0.5	0.7	0.9
5	6E-1	5E-1	4E-1	3E-1	3E-1
10	4E-1	3E-1	2E-1	2E-1	1E-1
20	2E-1	1E-1	9E-2	5E-2	3E-2
30	1E-1	7E-2	4E-2	3E-2	1E-2
60	3E-2	2E-2	1E-2	6E-3	3E-3

*   <m,f> stands for a lognormal distribution with median response time of m and an error factor of f.

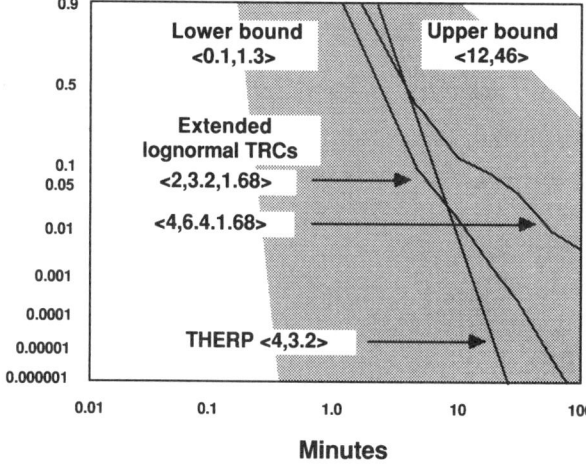

Figure 10-5. TRC system curves relative to the simulator data range.

Remaining Research Agenda

The TRC concept has gained much maturity in the last few years but still needs basic research. The concept will receive much more attention as more data comes in from simulators and, in particular, field data acquisition. The industry needs to review all of its incidents from the HRA and the TRC perspectives. Field data is needed to demonstrate the fidelity of simulator data, which can be collected on a routine basis. Also, more data is needed to provide realistic estimates of the uncertainties associated with TRCs.

Pattern analysis of the amount of data that can be anticipated over the next decade may resolve the burden issue (see Chapter 12 for more details), the effects of group dynamics on TRCs, the real influence of other influences, and the effects of a crew faced with tradeoffs between competing resources, which is a phenomenon evident in anecdotal reports of actual events. The TRC system described is a robust enough system to guide these data collection efforts and greatly enhance HRA.

Chapter 11
QUANTITATIVE ANALYSIS

The basic requirement from quantitative risk analysis placed on HRA is a set of estimates of the occurrence probabilities of all human failure events modeled in the PRA logic structures. In some industrial settings, there is enough data to make these estimates directly from data analysis. Since there is little such data in the nuclear industry (see Chapter 5), a technique to generate these estimates is needed.

Three kinds of techniques are popular in nuclear power plant HRAs: a task-analytical approach such as THERP (see Chapter 6), time reliability correlations (see Chapter 10), and a method that structures expert judgment of such events, such as the Success Likelihood Index Methodology (see Chapter 9).

Choosing one or a combination of techniques is a critical step in the HRA process shown in the schematic in Figure 8-1. The choice depends on the failure modes and mechanisms that dominate the human event as represented. The choice of the technique directs the analyst in identifying the characteristics of the sequence, task, and human behavior that need to be factored into the quantification. THERP and SLIM can accommodate and typically require numerous influences, whereas a TRC approach may only need the knowledge of the timing characteristics of the the sequence and the kind of behavior, e.g., rule-based that the event represents. After specifying the failure modes and mechanisms and the dominant influences on them, the technique allows a relatively straight-forward calculation of a point estimate of the probability of the modes or the total event. Some guidance relative to the quantitative uncertainty that should be associated with the estimate is also provided.

The Approach

Human failure events are quantified according to the following steps:

1. Human events are (usually) identified only in those classes indicated in Figure 9-1. This includes latent events as contributors to component unavailabilities and may include human contributors to initiators, if initiators are modeled rather than quantified directly from data. Slips may or may not be included in rule-based events. Recovery failures are usually attributed only to mistakes. Event specifics are identified by informal or formal interviews of operators and plant staff or observations in simulator exercises, whenever possible.

2. The events are represented and incorporated into the PRA structures either directly in a logic structure similar to the Operator Action Event Trees (OAETs — vonHerrmann, 1983) or indirectly as an adjunct to the PRA structures. An OAET is depicted in Figure 11-1 that represents the scenarios related to a

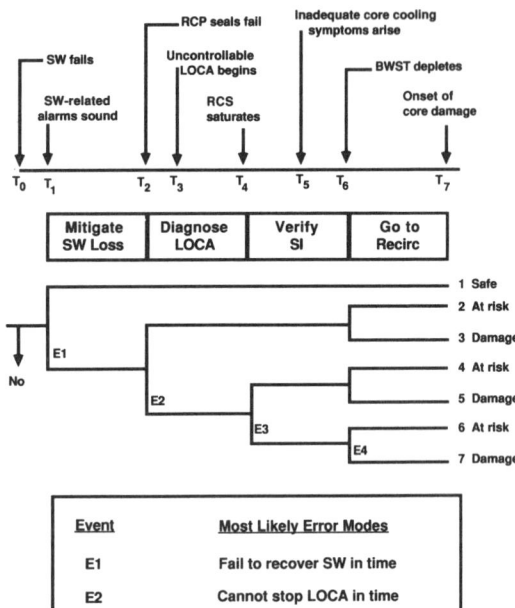

Figure 11-1. A sample operator action event tree.

reactor coolant pump seal LOCA. The figure also shows how a sequence timeline can be integrated with an OAET and the most likely human failure modes identified.

3. In the approach here, slips are modeled using a simplification of the THERP approach; mistakes are modeled using a combination of TRCs and SLIM.

4. Probabilities and their distributions are obtained according to the methods of the chosen technique (see Chapters 6 and 10).

The next sections outline the THERP-like approach to quantifying slips and the TRC system for quantifying mistakes. First, a strategy is presented that allows initial quantitative screening of human failure events.

Screening

The strategy for quantitative screening is depicted in Figure 11-2. The values for slips, taken from THERP, generally range one or two orders of magnitude around the value of 0.001. This value is used as a basic screening value for latent or human-induced initiator events. THERP also provides a value for the conditional

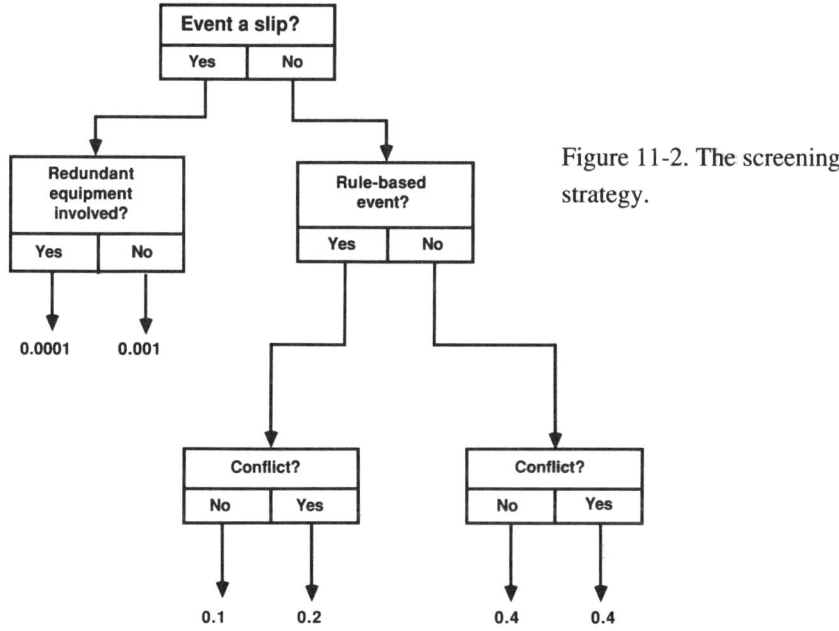

Figure 11-2. The screening strategy.

probability for a second error given the first. Assuming moderate dependency, this value is about 0.1. The product of these two values is used as a screening value for the human failure resulting in the failure of multiple (particularly redundant) components.

The time reliability correlations are used to quantify post-initiator events (as discussed later). The five minute value from each of the four basic curves is used to screen post-initiator events. The events are initially classified according to whether they are rule-based, rule-based with hesitancy, recovery, and recovery with hesitancy and the appropriate screening value is used.

Quantifying Slips

In Table 3-2, slips are meant to correlate to the failure modes of "misdetection" and "faulty actions". However, in a process plant, which is fitted with considerable instrumentation and annunciators, misdetection is not usually credible as a mode of failure. Thus in most cases, a slip is an action that is not as intended (Reason and Mycielska, 1982). In other words, the situation has been diagnosed and the decision or plan made, both successfully (otherwise there was a mistake). All that remains is for the person to do what was meant to be done. Slips occur all the time in everyday activities (and may be thought hilarious by the perpetrator if inconsequential). Slips also occur regularly in routine process plant activities. For example, a slip can occur when a maintenance person is returning a piece of equipment to service and fails to leave it in the proper state. Both auxiliary feedwater valves at TMI were left closed apparently by maintenance, which rendered a vital safety system unavailable. [This event was presumed a slip but its net effects were not typical.]

In the case of a slip, there is no unanticipated event to respond to and planning is specified by procedure, policy, or work order. It is fair to assume, then, that the main cognitive element is that of control, i.e., carrying out all steps of the plan. There are so many potential mechanisms of slips that THERP assumes that each step of a particular procedure is a candidate for a slip (i.e., an omission or commission) However, the likely candidates for slips in a control room seem to be limited to:

1. Stereotype capture, i.e., a familiar, more frequently performed task uses some of the same steps as the intended task, and in the first case of deviance, the operator slips into the old task rather than continuing what is called for in the required task,

2. Spatial reversals, e.g., two controls are adjacent and the wrong one is manipulated because of a lapse in attention (or a deviation from stereotype),

3. Time reversals, where the sequencing of manipulations is critical and the order is incorrect, again due to a lapse.

Assuming that the mechanisms of slips are limited to the above, only one instance of a slip per task is modeled, which is the first difference from THERP. A review of the estimates for task element probabilities in Chapter 20 of NUREG/CR-1278 (Swain and Guttmann, 1983) shows a range from 0.0003 to 0.003 for most of the types of situations relevant to process activities (see Table 5-2). Thus, a simple strategy is adopted of using the log-median of the bounds, 0.001, as a basic probability for any slip; 0.0001, or one order of magnitude lower, for a slip that leaves redundant equipment inoperable. This is the screening strategy mentioned previously. [Note that there is an argument that current equipment failure data already includes human failure contributions and a PRA need not opt to model latent slips at all.]

THERP, then, allows the basic estimate to be adjusted by so-called performance shaping factors (PSF) based on situational, task, or behavioral influences. For example in maintenance, there may be a check off policy and a PSF should give credit for this. The possibility of human engineering deficiencies can also be accounted for by a PSF, e.g., both valve operators for two trains are adjacent or reversed or otherwise could induce a reversal slip.

Quantitative Analysis 135

Table 20-22 of NUREG/CR-1278 suggests that the first factor should be 0.1 if the checking is independent of the manipulation. If a human engineering deficiency exists, then the second factor could be greater than 1.0, say, 5.0. The net result for the slip, then, would be:

event probability =
basic probability x factor$_1$ x factor$_2$ = 0.001 x 0.1x 5 = 0.0005. (1)

This simplified THERP approach may be used for slips whether they are latent, response, or recovery events. A more detailed approach was developed by Mike Lewis, then of INPO, for the Oconee PRA and is depicted in Figure 11-3. Here the minimum time between manipulations for a component is chosen. A manipulation is assumed to return the latent failure probability of the component to a basic probability, P_b. In the interim, the component may be subjected to surveillance or functionally tested. A factor, f_i, is assigned to each such act in the interim as:

$f_1 = 1$ (2)
$f_i = 0.1$ if only a surveillance is performed
    0.01 if a functional test is performed   for i≠1.

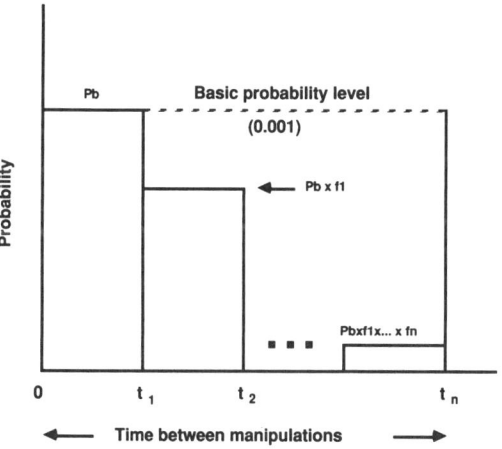

Figure 11-3. Process for estimating the probability of slips.

This function reduces the "current" failure probability. This factor reflects the assumption that functional tests both test for operability more reliably than surveillance and are performed as reliably. Assume there are n-1 such actions in the interim, which divides the interim into n pieces as indicated in the figure. The area under each step in the curve is:

$A_i = (t_i - t_{i-1}) \times P_b \times \prod f_j$ for j = 1 to i-1 and i = 1 to n. (3)

The net component failure probability, P_{comp}, is

$$P_{comp} = \sum A_i / t_n \quad \text{for } i = 1 \text{ to } n \tag{4}$$

where t_n is just the minimum time between manipulations.

The basic probability is the 0.001 value derived above multiplied by any performance shaping factor(s) as appropriate, f_{PSF}.

Multiple components enter the equation in one of two ways. (1) A group of components, e.g., a train of a multi-train system, may be manipulated by the same procedure set. A train (group) latent failure probability is then:

$$P_{train} = \sum P_{comp} \quad \text{summed over the number of components} \tag{5}$$
$$\approx n_{comp} \times P_{comp}.$$

(2) Some like-components in different trains may be manipulated together or systematically manipulated the same way over time. The prototypical situation is the calibration of redundant instruments, which is performed over a schedule with weekly separation. However, because of a miscalibration in the calibration equipment, both (or multiple) components are calibrated in the same, erroneous manner. In this case, the failure of interest is the latent failure of both (or all) components, and a dependency factor, ß, is used, borrowing from the dependency model of THERP. The probability of the failure of n like-components is then given by:

$$P(f_1 \text{ and } f_2 \text{ and } f_3 \ldots \text{ and } f_n) = Pr(f_1) \times Pr(f_2 | f_1) \ldots \times Pr(f_n | f_1, f_2, \ldots f_{n-1}) \tag{6}$$

where the subsequent probabilities are conditional probabilities.

Whenever all the failure events, f_i, are independent and identical, this relationship reduces to:

$$Pr(f^n) = Pr(f)^n \tag{7}$$

where the left side is the probability of the conjunction of n identical failure events and the right side is the product of the probability of one event n times.

Now,

$$Pr(f^n) = Pr(f)^n < Pr(f) \quad \text{since } Pr(f) \text{ is typically} < 1.0. \tag{8}$$

So,

$$\Pr(f^n) = \Pr(f) \times \beta \qquad (9)$$

where ß is a factor ≤ 1.0.

For the case of independent events, ß is $\Pr(f)^{n-1}$, which can lead to considerably low values for n>2. For example, for n=4 and Pr(f)=0.001 (a typical number from THERP), ß is 0.000000001. However, if Pr(f | g)=0.1 (or about the value for medium dependency in THERP) in the general formulation, then ß is 0.001, or six orders of magnitude higher than for the independent case. In the extreme case that all conditional events are completely dependent on the first event, ß is 1.0.

As a screening strategy beyond the simple version in Figure 11-3, let

$$P_\beta = P_b \times \beta \qquad (10)$$

where
ß is 0.1 for one other component,
    0.01 for two other components,
    0.001 for three other components, and
    0.0005 for four or more other components.

This formulation allows for medium dependency between three components and high dependency for four or additional components and gives some credit for the systematic error to be resolved over time, e.g., a new calibration instrument or a new person doing the calibration. Other formulations can be developed based on different dependency levels as the circumstance dictates.

Finally, a two train system would yield a latent failure probability of:

$$\begin{aligned} P_{sys} &= P_{train} \times \beta \\ &= 0.1 \times P_{train}. \end{aligned} \qquad (12)$$

This system gives a way to generically finely screen latent probabilities, which is usually all that is required in a PRA. Table 11-1 shows the fine screening procedure. Note that a train is assumed to consist of ten components.

THERP typically assumes error factors (i.e., measures of the uncertainty of the estimate) that range from 3 to 30. A log-median of 10 is used here; this implies that the order of magnitude of the point estimate so derived is accurate with a confidence of 90%, i.e., the order of magnitude of such estimates can be estimated reasonably well.

Table 11-1
Fine Screening Procedure for Latent Failures

Group of components	Probability	Type of testing
single	0.0001 0.00001	surveillance functional test
train	0.001 0.0001	surveillance functional test
like components/ different trains	0.00001 0.000001	surveillance functional test
system (two trains)	0.0001 0.00001	surveillance functional test

Quantifying Mistakes

A mistake results from one of the following failure modes: misdiagnosis, faulty decision, or faulty planning. These failure modes seem to matter most in a process environment when the process has deviated from normal or desired conditions, i.e., in an abnormal or emergency event. Since a mistake is often an error in decision making, it is natural to look to the literature of decision science for insights into these types of errors. Unfortunately, although there is a large literature in the decision sciences, most of it relates to investigating and identifying optimal decision strategies, i.e., how decisions ought to be made. When decision science addresses decision making as it is actually performed, the discussion is often qualitative and set in routine, everyday circumstances (Janis and Mann, 1977), which may have little direct bearing on response to off-normal process plant events.

It is known, however, that diagnosis and the decision making of a kind needed in off-normal response is influenced by available time, perceived time (or the pace or urgency of events), uncertainty, complexity, and goal conflict. These latter three factors are the primary sources of the phenomenon called burden (see Chapter 12). There have also been studies of simulated events (Beare, et al., 1982, for example) that show that the response performance of crews of operators to off-normal events as distributed in response time. These response times seem to fit lognormal distributions well (see Chapter 10).

Anecdotal reviews of actual events show that individuals involved in diagnosis and decision making often exhibit hesitancy. Hesitancy can be due to uncertainty in the conditions present, thus inhibiting or failing to allow for proper diagnosis

(Rogovin, 1980), or can be due to uncertainty as to which goals to pursue when one or more appear to conflict (Woods, 1982 and USNRC, 1985). The fact that conflict can be exhibited even when relatively clear procedures exist for these conditions shows that process plant operators are not automatons. However, this human flexibility implies fallibility and for this reason, diagnosis and decision making seems to be the critical and dominant risk-significant human behavior in an abnormal or emergency situation.

Because errors in diagnosis and decision making can have such serious consequences, nuclear power plants have attempted to eliminate or, at least, minimize the role of diagnosis and decision making by introducing so-called symptom-oriented emergency procedures. These procedures contain "rules" of the form:

IF symptoms $S_1, S_2, ..., S_n$ exist
THEN execute actions $A_1, A_2, ...,A_m$
AND do so without hesitation.

In most cases, the symptom set is kept small and each symptom is clearly indicated in the control room. The action set is one or two simple manipulations of controls also in the control room. It was the intention of this procedure style of small symptom set and simple action set that the necessity for on-the-spot diagnosis be all but eliminated and thereby reduce the error probability. While the use of simple IF-THEN rules is simpler and less ambiguous than the old event-oriented procedure, it is the mandate represented by the "and" clause in the rule that is not so clear cut. In fact, no such clause is written in the procedures; it is merely the intent of the rule that it be followed without hesitation. As a result, conflict or complexity or uncertainty from any source can introduce hesitancy in the heat of a real event. Combining the above concepts motivates the development of a model of process control mistakes that has the following characteristics:

1. Time is an independent variable; the probability of successful diagnosis or decision making (D&D) increases with available time. Available time is defined as the time from a clear indication for the need to act in a specified way until the time action would not produce the intended result (the point-of-no-return), minus the time it would take to implement the decision.

2. Crew effects are aggregated, i.e., the failure probability estimates are for an anticipated crew structure, such as defined in NUREG/CR-1278, Table 18-2, not for any individual.

3. Conflict, complexity, or other sources of burden are prime influences on the reliability of a D&D performance. This influence is explicitly factored. The presence of rules is also explicitly factored.

4. Other influences, such as the quality of procedures, the adequacy of the instrumentation and controls, and the adequacy of training, can be systematically factored into the quantification using SLIM, or some other subjective judgment approach.

5. Slips are not usually dominant when D&D is necessary, or when time is forgiving, or when multiple crew members are present to notice and correct slips. In cases in which slips are judged to be significant, the previously described THERP approach can be applied to misactions and the THERP annunciator model (Chapter 11 of NUREG/CR-12278) can be applied to misdetections, if postulated as important.

6. Early, incorrect action based on inadequate D&D (sometimes also referred to as a commission error) is assumed to be correctable, in principle. The likelihood of the error is subsumed in the TRC value. Its effects are not modeled unless they would change the course of the sequence radically.

As developed in previous chapters, mistakes are time-dependent stochastic process that are modeled by means of a family of lognormally distributed, time reliability correlations derived from the simulator work cited previously. There are two pairs of curves in the family. One pair is to be applied to response or rule-based mistakes, i.e., misdiagnosis or faulty planning or decision making when guided by the kind of rules specified above. One curve in this pair is to be used when hesitancy is not an important influence; the other curve is to be used when hesitancy is dominant. The second pair of curves is applied when on-the-spot, general diagnosis must be used to decide on a course of action in the absence of rules. This most often will be the case for recovery events. Again two curves are used to recognize hesitancy or its insignificance. Tables 10-8 through 10-11 listed several useful times for the four curves.

Each curve is mathematically characterized by a median response time, m, a measure of the central tendency of the response time distribution, and the error factor, f, a measure of uncertainty in the model. The median response time with its distribution, neglecting model uncertainty, results in lognormal TRCs. (See Chapters 4, 5, and 10 for the basis of the lognormal TRC family.) The curves for the four major failure categories are plotted on log-probability paper (Figure 11-4) to render the curves as straight lines. The recovery curves have the same median response time and the response curves also have identical medians. The hesitancy curves have the same error factors as can be seen by their parallel slopes, as do the non-hesitancy curves. The details of the derivation of the curves are included in Chapter 10. The recovery curve is a lognormal fit of the TRC which was used in

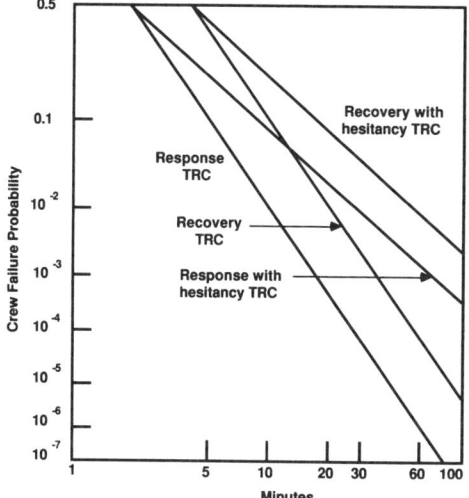

Figure 11-4. The basic time reliability correlations.

THERP (see Chapter 10 for its derivation) to apply to the diagnosis of choosing an appropriate procedure. THERP did not include a curve for rule-based responses.

Adding model uncertainty to the formulæ complicates the curves. The results are indicated in Tables 10-8 through 10-11 and are not plotted.

There are two major questions about this TRC system. First, the sixth characteristic of this model hints at the so-called commission problem: some diagnoses are so incorrect that reactor conditions can be significantly worsened if acted upon. TMI is often perceived as such a commission error, i.e., the crew thought the pressurizer was going solid and turned off emergency core cooling whereas, in actuality, the steam/water mix in the primary was producing a misleading indication and core cooling was needed desperately. The new functional, rule-based emergency procedures are designed to mitigate this very kind of mistake. They do so by providing rules at multiple levels of challenge to core cooling, any one of which acted upon would turn an incident around. The symptoms of these rules take precedence over any other symptoms and provide a redundancy to the normal diagnosis that would take place in a control room during an incident.

A generic calculation of the probability of such an unrecovered commission error can be represented by:

$$P = P1 \times P2 \times P3 \tag{12}$$

where
  P1 is the probability of an extended and significant commission error
  P2 is the probability that the ERGs do not cover the condition resulting from the error
  P3 is the probability of recovery by the SRO or other personnel.

In some 10,000 reactor scrams worldwide to date, only two involved misdiagnosis that led to core melt (including Chernobyl). So P1 can be estimated as 0.0002. If it is assumed that the ERGs reduce this estimate from one to three orders of magnitude, then P2 can be estimated as 0.01, the log-median. Finally, the SRO, as the supervisor of the crew, whose function is to stand back and monitor the safety status of the plant, should exhibit no more than low dependency with the other crew members. So P3 can be estimated as 0.05. The net result for unrecovered commissions in today's procedural climate is 0.0000001 (or 10^{-7}) per reactor year. This value is bounded by other failure modes.

The second problem with the TRC system is that lognormal CCDFs do not terminate at some low probability, although probabilities that are interpreted as human failure rates cannot meaningfully decrease without limit. However, the probability of unrecovered slips, even with multiple crew present, will serve as a lower threshold. This probability is on the order of 0.000001 using arguments similar to that used for the commission problem and a basic probability of 0.001 instead of 0.0002.

With these thresholds in mind, it has become standard practice to truncate failure probabilities generated by TRCs at 0.0001 or 0.00001 or to truncate the available time at about an hour, as an intermediate screening process. Following this convention, then only if a sequence event is risk significant and long term will lower numbers be considered.

The quantification system described so far depends on three factors: available time, the potential for hesitancy, and the presence of rules. It should be understood that other factors, such as the adequacy of the control room's instruments and controls, the applicability of procedures, the communication of the crew, etc., can influence the estimated probability of a human event. One way to factor in these influences is to use the Success Likelihood Index Methodology as an interpolation device. Using this device, each curve can be "adjusted" according to its assessed success likelihood index (see Chapter 10 for the adjustment and Chapter 9 for the SLIM process).

The net result is a procedure which allows for the quantification of human failure events of the mistake category. This procedure is outlined in Table 11-2 and allows the calculation of a mean probability of a mistake using the TRC system along with an assessed SLI.

ORCA

As noted in Chapter 8, the documentation required of the HRA by the PRA is extensive. The iterative nature of a PRA also means that the human failure event

Table 11-2
Quantification Procedure for Mistakes

No.	Step	Result
1	Specify human failure event, E	Qualitative analysis of E
2	Determine whether E involves a "rule"	k_C
3	Choose or calculate success likelihood index, based on specification of E	x
4	Calculate $2^{(1-2x)}$	k_I
5	Calculate $k_C \times k_I \times 4$	Median response time
6	Read one of the tables or use the ORCA algorithm	probability of E as a mean with 95th and 5th percentiles

database may have to be updated several times. As a rule-of-thumb, the documentation and more mundane data management tasks of an HRA make up about 20-40% of the analysis, depending the familiarity of the analyst with the plant being studied. To alleviate the routine tasks of an HRA, a software package has been developed. This HRA tool is called the Operator Reliability Calculation and Assessment (ORCA) code.

Computer Requirements

ORCA is designed to be used on an IBM PC/AT family of computers. ORCA was developed using Ashton-Tate's dBase III™ but is compiled so as to run without requiring dBase. The code can be installed on a hard disk or run from a floppy disk.

Application Scope

ORCA's basic purpose is to document all human failure events in a PRA. The main information kept on each event is shown in Tables 11-3 and 11-4. ORCA also has enough "intelligence" to ask the user for specific information and to work its way through an abbreviated taxonomy like that described in Chapter 9. Figure 11-5 shows the basic logic path that ORCA follows. The user is first directed by menus to name the database to work on and to name an event to add to, delete from, or edit in that database. If the add or edit options are chosen, ORCA asks the user to describe the human failure event as a slip or a mistake. Depending on this choice, ORCA asks for information sufficient to calculate a screening value for the event,

144   Human Reliability Analysis

Table 11-3
ORCA
Human Failure Event Record Sheet
- Mistake -

Event Designator   NDXOVERH   **Event Type**   Recovery

Event Description
The crew fails to realign equipment following recirculation hardware failures.

Option Information   **Screening Value**   4E-1

Rule-based?   no
Hesitancy?   no
SLI calculated?   yes
Standard TRC?   yes

Influences

		Rank	Normed-rank	Quality	Product
1.	Display adequacy	10	0.06	70	4.2
2.	Procedure adequacy	40	0.24	30	7.2
3.	Team effectiveness	20	0.12	80	9.6
4.	Communication effectiveness	10	0.06	80	4.8
5.	Workload	40	0.24	30	7.2
6.	Training adequacy	50	0.29	70	20.3
7.					
8.					
9.					
10.					
					53.3

SLI   0.53

Available time (min)   20

Mean Probability & Statistics   1E-2

    Lower bound   4E-4
    Upper bound   4E-2

    Median time (min)   4.0
    Error factor   3.2

Table 11-4
ORCA
Human Failure Event Record Sheet
- Slip -

Event Designator	NDTRNAX	**Event Type**	Latent

Event Description
The crew fails to restore train A equipment following maintenance.

Option Information		**Screening Value**	1E-3
Multiple equipment?		no	
Surveillance?		yes	
Functional test?		yes	
Other PSFs assigned?		no	

Performance Shaping Factors

1. Functional test   0.01

Mean Probability & Statistics	3E-5
Lower bound	1E-6
Upper bound	1E-4

which is done automatically. If the user wants to input more information, or perform a detailed calculation, or otherwise produce a final occurrence probability, ORCA queries for the appropriate information and the final, mean probability and its 5th and 95th percentiles are automatically produced.

Once initial results are obtained and reviewed, any changes in parameters or key descriptors can be input by using the event editing mode. The results are automatically propagated throughout the ORCA database and the HRA file is updated.

Reports Generated by ORCA

Several report formats are available that have been found useful on past PRAs. Table 11-3 shows a record sheet for a human failure event of the mistake category; Table 11-4 shows an ORCA report for a slip. ORCA also produces a summary listing of all events, one per line.

146   Human Reliability Analysis

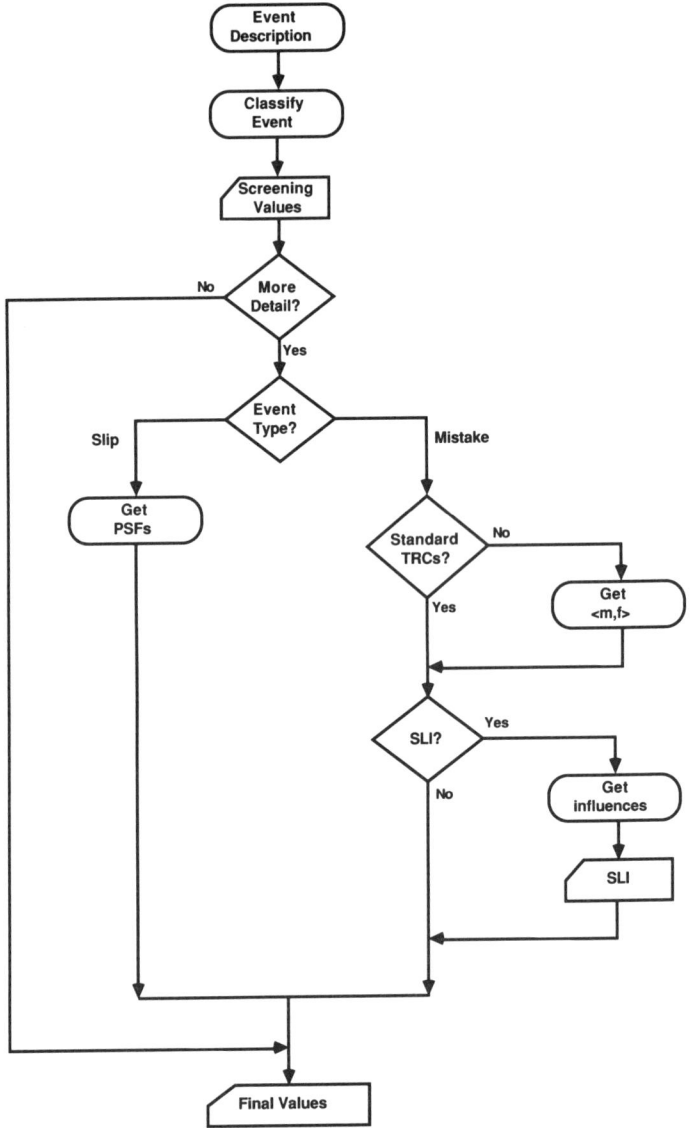

Figure 11-5. ORCA logic structure.

PART 3
Topics and Conclusions

Chapter 12
BURDEN

The role of a nuclear power plant operator was once described as 99% boredom with the other 1% interrupted by sheer terror. This is only partly an exaggeration. The part that is exaggerated is the claim that most of the time the operation of a nuclear plant is boring. A normal day shift routine consists of continual brush-fire fighting, from supervision of maintenance work orders to testing equipment to attending to regulatory items to managing incessant paperwork. A typical shift consists of four people; three are mandatory, five are a luxury and daily workload on a well-trained and motivated crew is considerable. Add to this workload the occasional unexpected event and the potential for overburdening a nuclear power plant crew is real and continual.

Psychologists have studied the finite capacities of people for years. Information processing is known to have distinct limits, e.g., the rule that only some seven ± 2 items can be retained in short-term memory (Miller, 1956) and the apparent single-channel capacity for conscious processing (Broadbent, 1958). The cognitive activities generally seem to be delimited, as exhibited by the reliability rate decrement as a person runs out of standard solutions (Wohl, 1982). Mental workload (Moray, 1979) is an active research topic in the psychological sciences. The mere need that a situation imposes on a person to make plans (Hess, 1987) or to make decisions (Janis and Mann, 1977) results in a challenge to precious, finite human resources.

External conditions also impose burdens on human capacities. A system fault or other event signature is complex to the degree of its interconnectedness or the redundancy of its parts or subevents (Henneman and Rouse, 1986). Such complexity is real but is also subjective, tied to cognitive limitations (Waller, 1982). There may even be "cases in which equipment complexity is such that human diagnosis will never be successful" (Wohl, 1983).

Finally, people have the apparently unique capability to act under uncertainty, always facing the dual questions: can I do it? and will something intervene? (Bell, 1979) Uncertainty represents a continuing relation between people and their world. However, uncertainty never enhances human performance (although clearly, some people handle uncertainty much better than others). For example, the time required to a decision increases with uncertainty surrounding the decision (Welford, 1968).

For this reason, most accidents in high-risk technologies involve human errors, but these errors are forced, induced by system complexity and unforgiving situations (Perrow, 1984). This chapter introduces a concept that may be behind forced errors: operator burden.

Sources of Burden

A review of incidents at B&W reactor plants led to the concept of operator burden (Black, 1987). Burden generally comes from time constraint. Either there is too little time to reliably perform an activity or the time rate of demands on the performer, the so-called pace of the events, is too great. However, other causal contributors to burden exist as well and are discussed in the next sections.

Diagnostic burden

The nuclear industry "discovered" diagnostic burden thanks to the TMI accident. Previous to TMI, emergency conditions were to be managed using event-oriented procedures, i.e., the event needed first to be diagnosed and the procedure appropriate to the event could then be implemented. TMI reminded us that diagnosis is a requirement in any off-normal incident and is not 100% reliable as a human process.

Much of the delay in proper response to the accident was due to the fact that the TMI operators first became aware of an indication that the core coolant level was rising and acted to prevent the overfill of the reactor vessel pressurizer. Unfortunately, this indication was false because the two-phase water conditions that resulted from the actual **loss** of coolant led the pressurizer level indicator to show an artificially high level. It was some 45 min until the operators properly diagnosed the event and began effective actions (NSAC, 1979).

The operators at TMI were confused by inconsistent instruments and fixated on the unusual and, in this unique case, misleading indication. Their actions were correct had their diagnosis been correct. Thus, part of the incident at TMI involved a forced mistake, in the HRA vernacular.

Another non-incidental factor was the fact that not only did the plant lose all main, i.e., normal, feedwater, but both trains of auxiliary feedwater (AFW) failed to actuate (apparently due to a common maintenance error prior to the accident). The failure of the AFW was noted but the actions that followed indicated that the crew did not believe in the total loss of feedwater, which was outside their experience base. Complexity also played a role at TMI. There were failures of main feedwater, failures of redundant auxiliary feedwater trains, a leak from the reactor coolant system's pilot-operated relief valve (PORV), and, later in the accident, the appearance of a hydrogen bubble.

TMI seems to show that there are three sources of burden on the operator during diagnosis: the credibility of ongoing events relative to operators' beliefs and experiences, the clearness in the indicated signature of the events, and the complexity or number of events at any one time. Other sequences that could occur across the spectrum of LWRs seem to offer the potential for diagnostic burden, e.g., confusion between a steam generator tube rupture and a small LOCA in PWRs, confusion induced by a loss of all dc power and thus all critical instrumentation in any LWR, multiplicity of events in the event signature under a major plant fire that disables the normal or emergency ac power in an LWR. Thus, diagnosis is influenced by the selectivity and belief capacities of people, and its reliability is a dominant issue in the reliability of emergency management.

Decision making burden

The loss of all feedwater at TMI also allowed the inadvertent discovery of a new option to cooling the reactor core. Without a source of secondary cooling, many PWRs can force-cool the reactor by injecting (or feeding) emergency core cooling system (ECCS) water and bleeding it from the system using controllable PORVs. This feed and bleed option can provide the necessary heat transfer in PWRs that are able to control the bleed operation. However, the option relocates the ultimate heat sink from outside the containment to inside and necessarily introduces radioactive water onto the containment floor (usually into a sump).

Putting heat and radioactive water where they are not desired obviously is not an inconsequential action. Further, there are a multiplicity of reasons for a total loss of feedwater (the secondary side cooling option) and many of these fault causes are readily repairable. This means that the operator is burdened not only with a decision to enact a normally undesirable option (feed and bleed) but has to negotiate a tradeoff between the restoration of the desired option and its alternative. All of this activity competes with attention and other resources of that are needed to combat the ongoing events and must be performed under the constraint of time imposed by an increasing temperature of a core without cooling. This scenario is further discussed in an analysis in Chapter 14.

Similarly, the potential decision in a BWR to manually actuate the automatic depressurization system under conditions which do not themselves actuate the system is another decision that may induce conflict. The actuation of the standby liquid control system during a BWR transient in which automatic scram failed is also a decision burdened by conflict, competing resources, and confusion.

There seem to be several sources of decisional burden: the fact of the need to decide itself, the potential for conflict between the alternatives, the fact that there are alternatives at all that may compete for finite human resources, the uncertainty in the effects of a decision, and the fact that an easy solution is the unpreferred solution and the preferred solution requires time and other resources to implement.

Command and control burden

The control room in a nuclear power plant is so extensively instrumented and full of equipment controls that most anticipated operator actions can be affected within the confines of that area.

However, some scenarios would take some or all of the crew from the friendly confines of the control room (see the subsequent example section). Then communication is not easily made, distance may separate operators from operators or controls from indicators, and the decision maker may not be able to first-hand supervise ongoing activities and see their results.

An example of the command and control problems that even routine activity may cause is the incident at the Sequoyah Nuclear Plant (Waage, 1982). There the unit operator dispatched an auxiliary unit operator to open two B loop residual heat removal (RHR) inlet valves and to verify that the interconnection valves between the RHR and the containment spray system were closed. The auxiliary unit operator arrived first at the interconnect valves and telephoned back to the unit operator who told him to open the "two valves". The auxiliary operator opened the (wrong) valves and proceeded to the inlet valves where the attendant telephone was inoperable. He proceeded to open the RHR valves which then opened a path from the high pressure reactor coolant water in the RHR system to the containment spray system. This created a LOCA with a loss of 40,000 gallons of water. Some forty-three minutes later the auxiliary operator, together with a second unit operator, pieced together the cause of the incident and the incident was terminated.

The major source of the burden induced by command and control is the remoteness of people and activities both from each other and in some scenarios from the control room.

Physiological burden

The concept of workload originally referred to the physiological demands that a job or task could place on its performer. As is the case for command and control, physiological burden is potentially minimized under normal activities which can be performed from the control room. However, remote, manual operations may be difficult to perform or may be conducted under adverse environmental conditions, e.g., inadequate lighting, heat, or tight quarters. In these scenarios, there is a source of burden due to the physical requirements of the intended actions.

Table 12-1 summarizes these general sources of operator burden.

Table 12-1
Sources of Operator Burden

Time constraint-related
    one action with a short available time
    multiple activities over a single duration

Diagnosis-related
    confusing indications
    credibility of events
    complexity of events or system

Decision making-related
    planning or decision making required
    conflict between an option and a normal intention
    competing resources

Command and control-related
    remoteness between people who need to coordinate
    remoteness of actions from control room
    distance between indications and controls

Physiology-related
    hostile environment

A Qualitative Example

Following the Brown's Ferry fire (USNRC, 1976), the NRC imposed several requirements by way of the Code of Federal Regulations (10CFR50, Appendix R) to avoid the perceived hazard to all LWRs from fires. The requirement is the assurance that one train of equipment can be used to bring a plant to hot shutdown. Cold shutdown must also be assured in 72 hours, but equipment repair or function restoration is allowed to meet this requirement. Many plants have implemented a policy of physically separating the electrical cables for shutdown equipment in different, redundant trains to help assure the availability of one train in the case of a fire-induced cable failure, as was the case for Brown's Ferry.

    One PWR recently was mandated to assure its proposed modifications met the fire requirements. This plant, referred to as Plant X, is an older unit and decided that the cost of cable separation was more than that of adding a dedicated diesel generator and the cabling from the diesel to a train of shutdown equipment. This dedicated shutdown system (DSS) was really a conglomeration of equipment located in various areas outside the control room with which operators could manually affect hot shutdown as prescribed.

Procedures were initially written that required the operators to disable the connection of the dedicated system equipment to the normal offsite power bus. The reason for this counterintuitive strategy shows the convolutions that power plants are forced into by regulators unconstrained by operational or risk criteria (or often, commonsense). Since the 10CFR50 regulation did not specify a particular fire or a frequency of risk from fire, no fix by plant personnel was ever sufficient. As soon as Plant X proposed a configuration of its DSS, the NRC inspector overviewing the plan "moved" the fire (this is not a problem unique to this utility and such a fire is sardonically referred to as a "smart" fire). As a result, the plant management felt that its only recourse was to prepare for the most unlikely and difficult fire scenario imposed on them, which could render ac power to all equipment in the plant untrustworthy. Thus, the potentially untrustworthy, offsite bus was to be abandoned, leaving operators **only** with the DSS option.

The likeliest scenario (identified by a fire risk analysis performed by the utility) involved the failure of one emergency bus that is violent enough to fail the adjacent redundant emergency bus (again a penalty of old technology in an old plant plus its lack of cable separation). In this case, the counterintuitive action of disabling offsite power would also be terribly incorrect in that it would eliminate the redundancy available to achieve hot shutdown that the operable offsite bus could provide. Disabling this source of ac is effectively permanent, since its restoration would take at least 8 hrs. Thus, risk might increase because of an improper regulatory requirement. Further, the plant operators, realized that the initial strategy was flawed and would have been placed in the position of having to make a decision by procedure against their better judgment.

This proceduralized but conflicting decision was not the only source of burden in the fire scenario. The equipment that together comprised the DSS was separated in at least five different regions in three facilities on three levels outside but near the main control room. If the fire disabled the two emergency busses or rendered the control room uninhabitable, then operators would have to affect plant shutdown remotely from the control room and from each other. The fire brigade that was assumed to be necessary to fight a major fire such as this would also leave, in many credible scenarios, only three operators to implement the DSS procedures, one for each major facility. As a result, the crew distribution would mean:

1. that many valves (on the order of 20 per operator) would have to be located and manipulated at various locations,

2. that access to these equipment would require special card keys or security guard keys that would have to be hurriedly found,

3. that coordination of activities, which was necessary, would have to take place using the plant paging system,

4. that the procedure might take the operator into areas lit only by emergency-lighting (if the recovery occurred at night), and

5. that the procedure might require the operators to go into areas filled with smoke or fire.

Further, many of the activities, if performed in proper sequence, required the operators to leave one area through its access, enter another, and return to the former more than once. Finally, all of this activity could, in a worst case, have to occur in 20-30 min. This is a significant physiological and command and control burden, as well as the decisional burden previously mentioned and the diagnostic burden of identifying the fire and its location as well as monitoring what equipment is available at any time in the event.

The modifications recommended to the plant from the HRA of the fire risk analysis included:

1. removing the requirement of disabling the offsite power to the DSS equipment except when that source is confirmed as unavailable,

2. eliminating some of the crossing of rooms by rearranging some of the procedural steps whose order was not vital,

3. relabeling all DSS instruments and controls in distinctive fashion to aid in their location, and

4. train yearly on the fire scenarios, with as much as possible of that training in the plant simulator.

Even with these fixes, the overall human reliability of this activity could not be assessed as very high. This example is one of many that show operator burden to be a real phenomenon that has potential risk significance. The next sections take rudimentary steps toward capturing this new HRA concern in some analytical manner.

156 Human Reliability Analysis

Sliding Time Windows

The intuitive notion of burden is "too much to do". The previous sections have shown more precisely what is "too much". This section attempts to point toward a potential measure of how much is "too much".

Figure 12-1 shows the 1985 loss of feedwater incident at Davis-Besse as a timeline of unlabelled events (a more detailed description is made in Chapter 14). Some of these events are actions taken by the plant operators; some were successful, some not; some were taken in the control room, some not. The other events represent changes in the status of specific plant equipment (failures and "successes") or in critical safety parameters, i.e., events that the operators would be expected to notice during the incident.

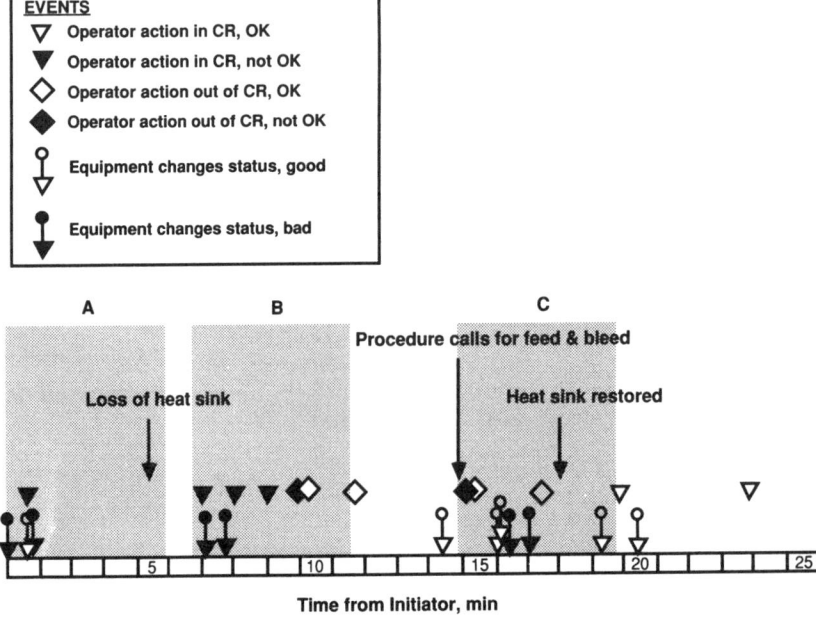

Figure 12-1. Sliding windows indicating burdened times.

By focusing on a (unspecified) duration—a window on the incident, if you will—regions of burdened activity seem to appear. Windows A, B, C seem to have been the three most significant regions of burden during the Davis-Besse incident.

<u>Window A</u>. This time duration saw the onset of the transient, a loss of all main feedwater to the steam generators. The reactor scrammed automatically (the only

success in the window) and the operator failed to manually actuate the auxiliary feedwater (AFW) system correctly. The major source of burden is diagnosis, at this time.

Window B. During this duration, several attempts to find alternative means to actuate AFW failed and much of this activity was conducted outside the control room. The reactor coolant system is heating up due to no heat sink to absorb the heat. Burden is due to factors relating to command and control and diagnosis (of the troubleshooting kind).

Window C. During this duration, the supervisor attempts to get the "last recourse" source of water to the steam generators, out of the control room, and succeeds. The RCS temperature reaches a maximum during this region. The conditions to go to the (undesired) option of feed and bleed, direct cooling of the core are reached at the beginning of the region. The operators are officially "stretching" procedures in this window, trying to affect the secondary side cooling restoration and deferring the decision to go to feed and bleed. Burden is due factors relating to decision making and command and control.

Intuitively, the degree of operator burden in the three windows increased over time because, partly, of the timed aspects of the phenomena of the incident. But also, it would seem, burden increased because its source changed from diagnosis, which is an expected, although non-routine operator activity, to command and control, a nonoptimal configuration with operators scattered over the plant, to, finally, the trauma of the decision to affect feed and bleed cooling or await restoration of secondary cooling.

Although no industry-wide analysis has been performed, it can be conjectured that all major nuclear incidents, say the precursors year by year (Minarick and Kukielka, 1982), pass through a burden spectrum similar to this one event. The more the "activity" per given time window, the more the burden. However, the nature of the burden may be even a more significant indicator of burden.

The last section discusses potential quantitative measures of burden, including the technique used in the HRA approach presented in Chapter 11.

Potential Quantitative Solutions

A quantitative indicator of burden must eventually exhibit a decrement in reliability, i.e., the probability of failure must increase whenever burden increases, with all other possible influences held constant. Repair troubleshooting has been observed to exhibit such a decrement in the rate of repair over time (Wohl, 1982). Figure 12-2 abstracts the situation. The decrease in the repair rate was conjectured to follow from the fact that many repairs can be affected by using standard routines that work

Figure 12-2. Decrement in reliability of maintenance tasks.

(Used with permission from *IEEE Transactions on Systems, Man, & Cybernetics*, Vol. SMC-12, No. 3, p. 243, May/June 1982. Copyright © 1982 IEEE.)

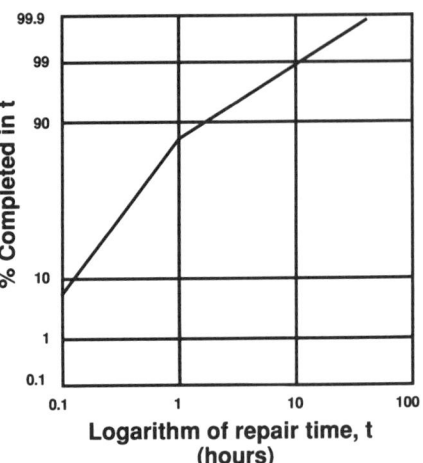

most of the time. Some repairs, however, can not be obtained using these normal heuristics and the repair requirements must be worked out using first principles or other knowledge-intensive strategies. The exhibited decrement in the repair rate thus seems to merely reflect the ultimate switching over to slower, more difficult cognitive processes. Unreliability, in effect then, is the result of a mismatch between process time and available time. This was the reasoning behind using the different TRCs for hesitancy in the system described in Chapter 11. This reliability decrement is essentially due to the change in requirements on behavior, in the example, a change from using patterns of past experience to using first-principles problem solving.

Decrement due to a change in behavior type using the TRC approach

The approach described earlier in effect doubled the error factor of the underlying response density from 3.2 to 6.4. In fact, a more general approach can be developed, as was in Chapter 10, in which:

$$f = 3.2\ e^{cz+d} \text{ and } z \in [0,1]. \tag{1}$$

In this way, the parameter z acts as the "degree" of burden measured from worst equal to 0 to best equal to 1 (similar to a success likelihood index). The result is a multiplier on the basic error factor of 3.2. The described approach effectively took z as $-d/c$ when there was no source of hesitancy (or burden) and $(\ln 2 - d)/c$ when sources of hesitancy were present. [For extreme cases of burden, the value for z may be adjusted even closer to unity.]

Another approach to burden has been used in other contexts. If the sources of burden are present in a postulated event, then it is conceivable that some non-zero threshold probability of failure would persist over the duration of the event, i.e., success probability is asymptotically less than unity. Because a TRC is the CCDF of a response probability density, the success probability can only approach unity asymptotically. Some HRA developers have truncated the failure probabilities obtained by a TRC above an artificial but desired level, such as 10-4 or 10-5 (Hall, et al., 1982). This practice, of course, accomplishes the intended purpose, but is not very mathematically elegant.

An alternative is just as ad hoc but somewhat more elegant. The times available in most risk-relevant sequences seldom are more than an hour. (This is due to the fact that an uncovered core for but a few moments leads to significant core damage and core uncovery is on the order of 30-60 min for most sequences.) A way to assure "high enough" failure probabilities for most likely available times is to calibrate the TRC so that it obtains the desired truncation value only after an hour available time. This can be easily done with the lognormal family of distributions by increasing either or both of the median time or the error factor.

How many events are too many?

The above technique allows for a way to quantitatively reflect burden that is due to a change in required behavior. Another variation arises from the notion that is often called (mental) workload (Moray, et al., 1979). This concept asks—how many events are too many?

If multiple events occur nearly cotemporaneously, they compete for attention and jointly use other resources, such as diagnosis or multiple crew members. Multiple indications for a single event may also be so different as to induce operators into performing responses to the indications in such a serial fashion. Of course, multiple indications of a single event may instead be reinforcing, i.e., they may convey the same overall message but merely with differing modalities or signatures. For example, the feedwater pump trip alarm and the rapid descent of the steam generator level both, and particularly together, indicate the loss of main feedwater cooling. Such a multiple alarm signature would likely be treated as a single indication by an experienced crew and not increase burden over a one alarm signature. (In fact, if one of the indications failed to arise, burden would probably increase due to an unmet expectation.)

The THERP nominal diagnosis model provided a second TRC to be used for a second different cue (whether of one or more events), when cues occur near in time. This concept can be more rigorously accounted for by recognizing that the

160   Human Reliability Analysis

THERP technique is essentially a convolution of distributions (which in this case are the probabilistic complements of TRCs).

Suppose three distinct signatures arrive in a control room essentially at once. Suppose the response to each requires the decision-making efforts of the shift supervisor and he or she can only deal with one at a time. Suppose further that the time from the first indication to some undesired plant condition, such as core damage, is t. Let t_1 be the duration it takes to decide upon and affect a response strategy for the first event, and t_2 and t_3 be analogously defined for the other two events. Then success is obtained only if:

$$t_1 + t_2 + t_3 \leq t \qquad (2)$$

If it is assumed that the response to each event alone is the recovery TRC, then the probability of non-response to all three events (and thus the third, since the decisions are made serially) is the convolution of the TRC three times. Denote such a convolution as T^3, where **T** represents the random variable of response time to a single event. Table 12-2 shows the effects of the number of convolutions, i.e., events, on the overall failure probability vs. time, for the basic rule-based TRC, a lognormal <2, 3.2> curve, and the basic recovery TRC, a <4, 3.2> curve. Figure 12-3 graphically depicts **T**, T^2, and T^3, in which T is the <4, 3.2> recovery TRC.

Table 12-2
Effects of Multiple Convolutions of the Base TRC

Available time, min	Number of convolutions		
	1	2	3
Rule-based TRC			
10	0.01	0.06	0.2
20	0.0003	0.002	0.009
30	0.0003	0.0003	0.0004
Recovery TRC			
10	0.1	0.4	0.8
20	0.01	0.06	0.2
30	0.002	0.01	0.04
60	0.0003	0.0003	0.0004

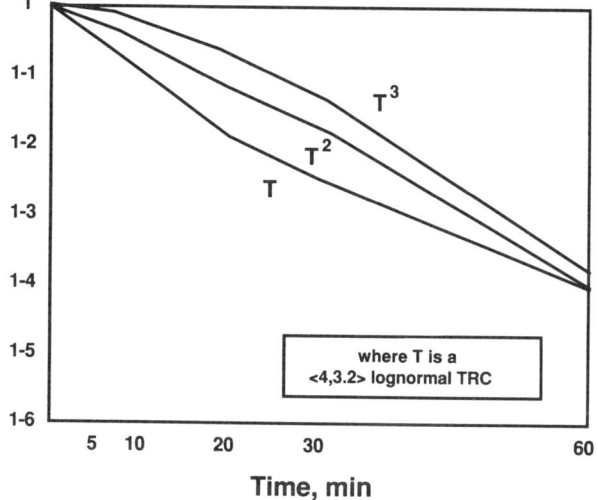

Figure 12-3. Convolving a TRC, 1, 2, and 3 times.

One way to quantitatively measure burden is to convolve the underlying (basic) TRC n times and note the available time that is required to achieve some threshold probability of failure. Figure 12-4 depicts this process for **T** and **T³**. Table 12-3 lists the required available time to reach three thresholds—0.5, 0.1, and 0.01—for rule-based and recovery TRC convolutions in which n varies from 1 to 3. (Note that when n=1, there is no convolution and the result is the underlying curve.)

In experimenting with the convolutions of the recovery TRC, an interesting correlation was noted as exhibited in Table 12-4 and graphed in Figure 12-5. The addition of conflict as described in the previous section—by doubling the

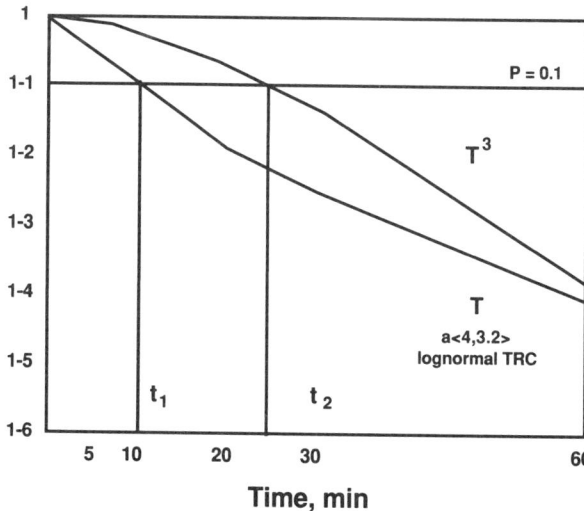

Figure 12-4. The available time required to achieve a threshold probability of failure.

Table 12-3
Available Times (min) to Achieve Various Failure Thresholds

Threshold probability	Number of convolutions		
	1	2	3
Rule-based			
0.5	2.0	4.5	7.0
0.1	5.0	8.8	12.2
0.01	10.3	15.3	19.6
Recovery			
0.5	4.0	9.0	14.0
0.1	10.0	17.5	24.5
0.01	20.6	30.6	53.0

Table 12-4
Correlation of Convolution to Conflict

Available time min	Recovery TRC		Convoluted TRC	
	no hesitancy	hesitancy	1	3
10	0.1	0.2	0.1	0.8
20	0.02	0.09	0.01	0.2
30	0.005	0.04	0.002	0.04
60	0.0003	0.01	0.0003	0.0004

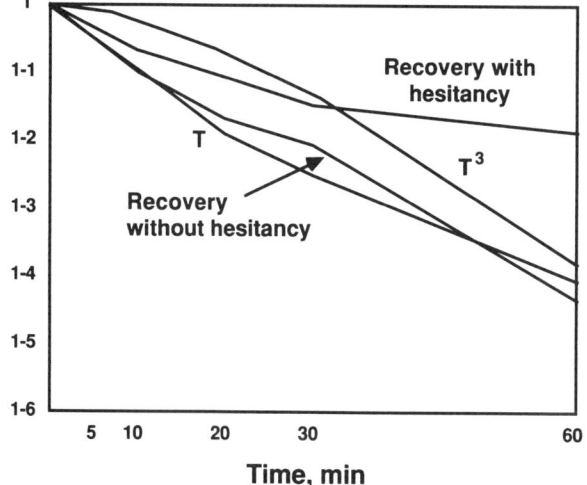

Figure 12-5. Correlation of conflict to convolution.

error factor—is nearly equivalent over 30 min to convolving the originating recovery TRC three times, i.e., n is 3. This equivalence seems to mean that conflict, a behavior-oriented concept of burden, is related to the numerical degree of complexity associated with diagnostic or decisional burden. That is, conflict, which probably has degrees associated with it, is equivalent to the resolution of how many is too many.

Whether this is coincidence or has some deep meaning is speculative at this time. However, the truth of the correlation would mean that the threshold probability, which is related to how many events are too many, would then be related to the error factor of the underlying distribution, which in turn would be related to the number of distinct events presented in an accident scenario. Continuing the development of these concepts could provide a procedure for reviewing past significant events to determine where burden was likely to have occurred and to investigate the subsequent resulting actions. It also suggests an approach to review postulated accident sequences, such as in a PRA. A time window could be established which provides an "acceptable" probability of response for a given number of serial events and this time window could be moved along the sequence timeline as suggested in the previous section. Areas where the number of events exceeds the acceptable could be highlighted for further study.

Chapter 13
GARONA RECOVERY ANALYSIS: A BWR EXAMPLE*

The Santa Maria de Garoña nuclear power plant, near Miranda, Spain, is a 440 MWe BWR plant that has been operational for about 15 years. In 1984, Nuclenor, their corporate plant operator, decided to perform a level 1 probabilistic risk assessment (PRA) for Garoña. This chapter discusses the human reliability analysis (HRA), in particular the recovery analysis, for that PRA.

Garoña has a one-train, turbine-driven high pressure coolant injection system (HPCI) and an isolation condenser (IC) for high pressure reactor level and pressure control in case the feedwater system or main condenser are not available. A two-train low pressure coolant injection system (LPCI), the core spray system (CSS), and the reactor shutdown coolant system (RSCS) provide low pressure reactor level and pressure control. The automatic depressurization system (ADS) provides a quick way to depressurize the reactor pressure vessel to use the low pressure systems. LPCI and the drywell ventilation system provide containment cooling (LPCI for the torus) and containment pressure control. Typically, short-term core cooling is implemented at high pressure and long-term cooling at low pressure. Depressurization can be affected manually and long-term cooling must be affected manually.

The core damage risk profile, which was developed during the course of the PRA for Garoña, consisted of ATWS sequences, loss of station ac power sequences, and a variety of others. From this perspective, Garoña is much like BWRs in the United States. One reason that sequences other than ATWS and LOSP contributed only slightly to core damage risk was the highly assessed reliability of the Garoña operators in recovering from these transient conditions. These sequences are often identified by the nomenclature originally developed in WASH 1400 (USNRC, 1975) as sequence TW.

The operational details of the recovery from the transient/long-term cooling sequences are developed in detail in the rest of this chapter. These sequences are depicted in the event tree of Figure 13-1. Note that this structure is a synopsis of several event trees developed in the Garoña PRA and is not fully representative of any of them.

---

* The work behind this case study was performed by the co-authors and E. P. Collins of SAIC, and by René A. Fernández of Nuclenor, the utility performing the PRA.

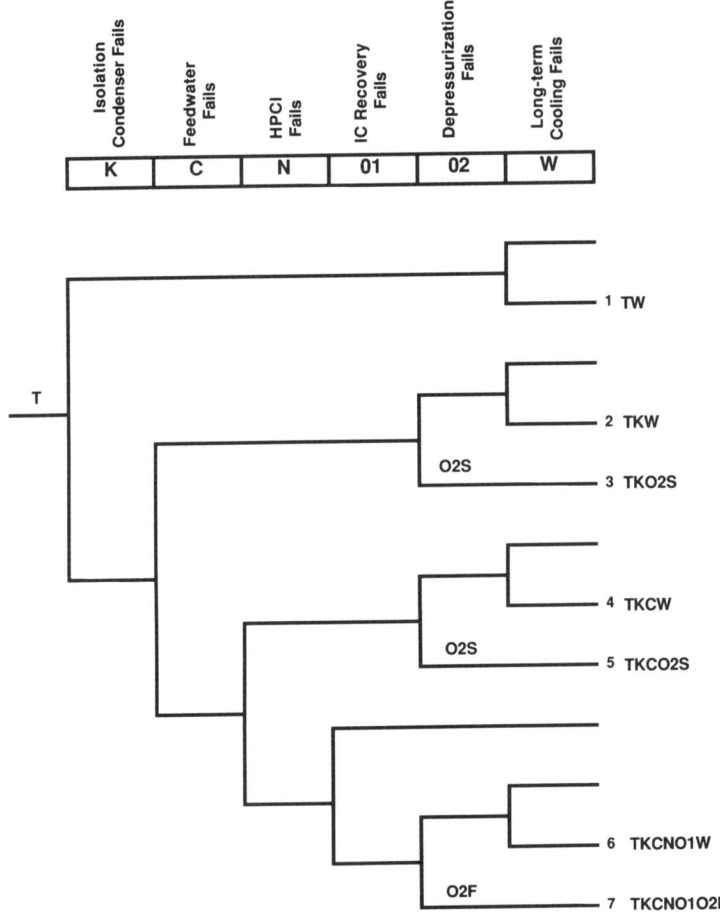

Figure 13-1. Transient event tree for Garaña plant.

The Transient Sequences and Human Reliability Questions

Several systems can be used to recover from a transient, T, that does not involve the loss of external ac power. If the main condenser is inoperable or the reactor vessel is otherwise isolated, the IC is the preferred mitigation option. The IC can provide cooling and maintain reactor level using natural convection, although it requires manual initiation of condenser makeup in an hour or so. In lieu of the IC, feedwater (if still available) or HPCI can maintain level control at high pressure. Some reactor pressure relief and torus cooling will also be required if the main and isolation condensers are unavailable. Long-term cooling also requires reactor depressurization. If all of the high pressure systems fail early in the sequence, then manual initiation of ADS can lower the pressure to allow the use of the low pressure systems: LPCI, CSS, and RSCS. Torus cooling requires using LPCI; otherwise, RSCS is the preferred low pressure system.

The event tree in Figure 13-1 shows two operator failures: O1, the failure to recover IC operability; and O2, the failure to depressurize the reactor pressure vessel. O1 was chosen because the IC is the preferred system in transient mitigation and its likely failure modes are recoverable. O2 was chosen because in all cases except successful IC, the RPV must be depressurized at some time to implement safe shutdown. HPCI and feedwater were considered unrecoverable (in fact, the Garoña operators suspect the reliability of the HPCI). All long-term cooling requires manual actuation, but this action was considered to be coupled to the depressurization that is also required. Thus, O1 and O2 are the only human failure events included in the event tree. Latent human failure events were modeled in the system fault trees but did not turn out to be important in sequence cutsets.

Several other simplifying assumptions lead to the sequence types shown on the tree. The assumptions are:

1. RPV integrity is assumed, i.e., transient-induced LOCAs were covered elsewhere.

2. IC is long-term operable except that it requires makeup to the condenser in an hour.

3. Feedwater and HPCI were assumed non-recoverable; anyway, IC recovery would take precedence over the recovery of the other two.

4. As long as reactor vessel level is maintained, the operators need only to depressurize slowly to the low pressure systems to achieve long-term cooling.

5. If level control is lost, the operators need to quickly depressurize using the ADS valves. This action is manual since ADS requires LOCA conditions to initiate automatically.

Seven sequence types result. These are:

1. TW - this is the specific sequence that is analogous to the WASH 1400 sequence in which the only failure following a transient is a crew failure to initiate one of several systems or hardware failures of all systems.

2. TKW - sequences in which the IC is not available and long-term cooling fails due to hardware.

3. TKO2S - same sequences as 2, but with failure of the operators to depressurize or otherwise initiate long-term cooling.

4. TKCW - same as 2, but feedwater is not available.

5. TKCO2S - same as 3, but feedwater is not available.

6. TKCNO1W - failure of all high pressure systems and long-term cooling hardware failures.

7. TKCNO1O2F - failure of all high pressure systems and long-term cooling due to human failure.

From a review of the human interactions with these sequences, the key HRA questions appeared to be:

1. Is TW really an issue from the perspective of human reliability?

2. Are sequences 3 and 5 different with respect the human reliability, i.e., does O2S differ on different branches?

3. Is sequence 7 valid, i.e., are O1*O2 or O1*not-O2 possible when their operational and human factors are considered (note: not-O2 is a successful action)?

4. How can O1 and O2 be quantified, taking into account the BWR Owners Group (BWROG) emergency procedures (EPGs), sequence characteristics, and other influences on human reliability?

Each of these questions is addressed individually in the following sections.

TW - An Old Concern

WASH 1400 assessed there to be no credible human failures in the recovery of the sequence that it denoted as TW. This strategy probably should have been adhered to. Instead, human reliability analysis gained enough sophistication to become aware of so-called cognitive failures, i.e., the failure of people in diagnosing a problem and deciding how to recover from it.

For several years, arguments flared over the reliability of a full crew of operators plus any support emergency staff in remembering to initiate one of typically 3-5 systems to affect long-term cooling. The assumed transient was benign enough to only cause reactor trip, and maybe RPV isolation or feedwater loss, but nothing else. In the most benign cases, reactor inventory makeup would not be required for tens of hours. In the most severe case, for example when using the IC at Garoña, makeup might be required in one hour. In any case, it is difficult to infer credible failure modes that would survive the crew and resource redundancy present in a BWR.

However, typical screening values for long-term human failure events, i.e., frequencies as low as one failure in 10,000, were not low enough to eliminate this event, as described, from risk consideration. Also, the task analysis techniques of WASH 1400 and, later, NUREG/CR-1278 (Swain and Guttmann, 1983), did not seem to account for the cognitive aspects of the event and were, at the time, judged inappropriate.

The solution (circa 1984) was to declare this kind of event as incredible. The incredulity arose much from the fact that the new BWROG emergency procedure guidelines are specifically designed to handle such an event. For Garoña, if reactor vessel level is challenged, then the Garoña plant emergency procedure POE-01 directs the operators to try to use, in the following order of preference, the feedwater/condensate systems, the control rod drive system, HPCI, CSS, and then LPCI. If only reactor pressure is challenged, then emergency procedure POE-02 directs the operators to use the isolation condenser. The procedure for implementing IC in turn calls for its long-term makeup, then PORV/RV relief, HPCI, and three other systems. If neither level nor pressure are challenged, then the core is not vulnerable, and the sequence is not risk significant. As a result of interviews and procedure talk-throughs, it was determined that the Garoña operators were thoroughly versed on these resources and the implications of the entry conditions to the EPG/POE system. For this reason, this event was considered negligible in the Garoña HRA.

Different Conditions, Same Action

The event, O2S, represented the failure of operators to slowly depressurize the reactor vessel following a transient in which either feedwater or HPCI successfully maintained vessel level. So long as level is maintained, this action does not need to occur for hours. The long-term cooling requirements are unaffected by the means that level is maintained. Because of these considerations and the fact that the POE procedure system is function-based, the action was considered generic and applied to both sequence 3 and 5. As in TW, O2S was considered highly reliable, although, in this case, the above mentioned screening value (0.0001) was sufficiently low enough that these sequences were not risk significant.

Event Interpretation

On the surface, sequence 7 would seem to be invalid because it includes the conjunction of two operator failure events. That is, it would seem likely that

Pr(O1 * not-O2)=0, or
O1 * O2=O1 in Boolean logic terms,

which means that sequence 6 disappears and sequence type 7 is TKCNO1.

However, there are details, i.e., cutsets, of this sequence class that need closer examination. Since T, C, N, are assumed not to be concerns of the recovery analysis, this event reduces to KO1O2. Event, K can be partitioned as follows:

K = KER + KEN + KLO + KLH

where

KER is the recoverable failures of the IC early in the incident, i.e., essentially coincident with T.

KEN is the non-recoverable failures of the IC early.

KLO is the failure of the operators to initiate makeup to the IC in an hour, i.e., human failures of the IC late mission.

KLH is the hardware failures of the IC late.

Based on operator information, it was assumed that, upon late failure of the IC, the operators would switch to an alternative to the IC rather than attempt its recovery. Recoverable failures of the IC were electrical or control faults in the motor-operated valves of the IC and all other failures were considered non-recoverable. Thus, the partitioning of K above fairly represents the plant and its operational strategies.

Table 13-1 shows how the elements of K can be interpreted together with O1 and O2 or not-O2. The first column is the Boolean representation of the combination. The second column is the representation after the event has been interpreted relative to its human reliability aspects. The third column summarizes the interpretation. Details are provided next.

KER*O1*O2F. The failure of the IC is due to recoverable electrical faults to motor-operated valves. The operators would have indirect indication on a control room panel that the valve are not in proper position. There would be 10-15 minutes until the torus heated up to 43°C, a time in which the operators would attempt to manipulate the valves. At the specified temperature, the crew is supposed to begin torus cooling and prepare for long-term cooling, which would not be required for another 30-60 minutes.

The Garoña operators exhibited a high preference for using the IC when the main condenser and feedwater are unavailable, which suggests that IC recovery will be attempted. However, the implementation of the BWROG EPG/POE system suggests that the torus temperature will be enough of a leading cue to make event O2F independent from O1. This is one way in which the symptom-based procedures were assessed to have an impact on the HRA.

The "likely" failure scenario in this representation is that the operators do not attain IC recovery before switching to torus cooling and long-term cooling requirements. The latter action fails for insufficient time.

Table 13-1
Event Interpretation

Combination	Representation	Interpretation
1. KER*O1*O2F	KER*O1*O2F	IC recovery fails because recovery action takes too much time; torus temperature, 43°C, cues long-term cooling but fails due to time running out or too few resources
2. KER*O1*not-O2F	KER*O1	same as above but long-term cooling succeeds
3. KEN*O1*O2F	KEN*O2F	long-term cooling fails because of too much concern for IC recovery; run out of time
4. KEN*O1*not-O2F	KEN	IC cannot be recovered but successfully implement long-term cooling
5. KLO*O1*O2F	KLO	all credible or probable modes of failure of actuation of IC long-term would fail any other action
6. KLO*O1*not-O2F fails	not valid	as in above failure to actuate IC long-term other actions
7. KLH*O1*O2F	KLH*O2F	assume late failure of IC directs operators to implement long-term cooling
8. KLH*O1*not-O2F	KLH	same as above but long-term cooling succeeds

KER*O1*not-O2F. This scenario is the same as the immediately prior one except that long-term cooling, in particular torus cooling, is achieved in time.

KEN*O1*O2F. In this scenario, IC cannot be recovered. Either this effort interferes with the torus cooling activities or time runs out and O2F occurs. Failure O1*O2F is reduced simply to O2F.

KEN*O1*not-O2F. There is no human failure in this case, since torus cooling and long-term cooling succeed and IC recovery cannot be achieved and thus cannot fail.

KLO*O1*O2F. The IC must be replenished with water in about an hour. This requires the opening of a single motor-operated valve from the control room. A failure mode analysis of this action is presented in Table 13-2. The probable failure modes or mechanisms seem to preclude the success of long-term cooling. IC recovery is assumed integral in the event KLO. Thus, KLO*O1*O2F reduces to KLO.

Table 13-2
Failure Modes of KLO

Mode	Mechanism	Likelihood
Slip	valve 1301-3 or 4 could be manipulated instead of 1301-10	negligible
	operators might omit the step in the IC procedure to initiate condenser makeup	highly recoverable; negligible
Mistake	since the critical reactor parameters would not degrade until some time after the IC ran out of water, the focus on the critical parameters might mask the IC problem and the crew run out of time	time-dependent

KLO*O1*not-O2F. Since O1 and O2F are totally coupled with KLO, this scenario is not possible.

KLH*O1*O2F. Operator information indicates that, since late IC failure would occur with the reactor near hot shutdown anyway, the operators would not bother with recovering IC. This representation reduces to KLH*O2F. O2F can occur due to time running out.

KLH*O1*not-O2F. This is the same as the prior scenario except long-term cooling succeeds. The representation of this scenario is KLH.

This example from the Garoña HRA demonstrates the interplay between an event tree representation of a scenario, which is function and system oriented, and the human reliability interpretation of the scenario, which is based on a deeper, i.e., cutset, representation. This interplay must be occur whether represented explicitly in the development of the event trees as was partially done for the Garoña PRA or later when the cutsets are better known.

Garoña Example 173

A Photo Survey

The study of the Garoña plant allowed the introduction of a qualitative analysis alternative to simulating sequences, a photographic survey or walkthrough of the candidate dominate accident sequences. Garoña does not yet have a simulator but the plant personnel were willing to allow an off-shift crew of operators to walkthrough the sequences in the control room, while the plant was at full power operation (a virtual impossibility in the US). One of the Garoña sequences is discussed and four of the 28 photographs taken during the survey are used as illustration.

The sequence of interest is a trip of the reactor due to the tripping of the feedwater system. The prevention of core damage can be accomplished by using the isolation condenser (IC) as an alternative feedwater system, or resorting to the high pressure coolant injection system (HPCI), or depressurizing the reactor and using the low pressure coolant injection system (LPCI) or the core spray system (CS). This is sequence seven of Figure 13-1, a TKCNW sequence, where T is the loss of feedwater transient, K is the failure of IC, C is the loss of feedwater which is the transient itself, N is the failure of HPCI, and W is the failure of both long-term cooling systems. The key issue in this sequence is whether the operators would attempt to salvage some source of high pressure water, in particular IC, or would depressurize to use some low pressure system.

Table 13-3 itemizes the actions that an operator would likely take under the conditions at hand. Four of these steps and their accompanying photographs were considered critical to the sequence's human reliability evaluation and are discussed next.

Action 1 (from Table 13-3). The feedwater trip would be alarmed on Panel 906-9, 22A, as low flow on the feedpumps and low suction pressure on the condensate. The reactor would trip within 8 seconds from the feedwater trip. The initial focus of the operators would be on the impending trip.

Figure 13-2 shows a Garoña operator at the middle of the L-shaped control board pointing to the alarms that indicate the status of the feedwater system, which would be the initial cue for the sequence.

Action 8. The emergency procedure guidelines would first direct the operator to assuring the shutting down of the reactor, which is assumed to be successful and automatically accomplished in this sequence. The first critical safety parameters to monitor are the reactor vessel level, which would rapidly decrease, and the reactor coolant pressure which would rapidly increase. The POEs would direct the operator to prepare to use the IC, HPCI, and the low pressure systems in that order and the IC would be of first concern. (The Garoña operators stated that the IC is not only

Table 13-3
Operator Actions in Sequence TKCNW

1. Alarms on panel 906-9, 22A "low flow on feedpumps A, B, and C" and "low suction pressure on condensate" sound, indicating the feedwater transient. Operators acknowledge and station themselves.

2. The reactor operator attempts to restart the feedwater pumps (assumed in this sequence to be unsuccessful).

3. Alarm on panel 905-24, 28A "low level reactor scram A & B" would indicate that the reactor has scrammed. RO verifies the source of the scram using the high and low scale reactor level meters on Panel 905.

4. RO initiates backup manual scram by pushing the scram buttons.

5. RO verifies automatic tripping of the recirculation pumps.

6. RO monitors the automatic isolation via the main steam isolation valves, indicated on mimic panel 903-22, 23 and by the valve closure lights on panel 903-32, 33.

7. The assistant senior reactor operator notes that several relief valves open, alarmed on panel 903-4, 26A. ASRO would check which had opened by monitoring the valve pressure meters on panel 903-5 or 903-6.

8. The ASRO would attempt to control reactor vessel pressure by removing heat using the isolation condensor. The sequence assumes that the IC is unavailable. One recovery is to open the reactor inlet valve from the IC via the switch on panel 903-10. Failure of the IC is indicated by alarm on panel 903-4, 26A "isolation condensor valves off normal".

9. Meters and warning lights on panel 903-15 indicate that high pressure coolant injection did not start automatically. The ASRO turns attention to manually starting the HPCI pump.

10. Failures of IC and HPCI, accompanying the feedwater trip, leave the crew with no high pressure coolant source. The assistant senior reactor operator would verify the operation of the low pressure systems by starting the low pressure coolant injection pumps using switches on panel 903-11 and 12 and the core spray pumps using switches on panel 903-9 and 12. If either system were to work, the ASRO would depressurize and use the working system. LPCI is the preferred system when both are available.

The sequence may terminate, as described, before depressurization, if the failures of the low pressure systems, event W, involve start failures. If not, depressurization may be affected before W occurs. The operators continued the walkthrough just to demonstrate the depressurization and low pressure system operation actions.

Table 13-3, Cont.

---

11. Operators check rate of depressurization indirectly by monitoring the reactor level meter on panel 903-6. Depressurization is attempted using two or more relief valve/safety relief valve switches on panel 903-11 and 12, 903-9 and 10.

12. The operator checks LPCI valve alignment on panel 903-11 and 12 and the flow of emergency service water through the LPCI heat exchangers using meters on panel 903-5.

13. Reactor level is monitored on the meter on panel 903-6 to help guide the control of flow through the LPCI pumps during long-term cooling.

Figure 13-2. Operator recognizes transient type.

their logical first alternative but expressed high confidence that the system, being simple and nearly passive, would work and "solve" this sequence.) There are two likely failure causes of the IC that would be relatively easy to recover from and IC failure would induce the crew to consider these recoveries.

Figure 13-3 shows an operator attempting to open the valves between the IC and the reactor vessel in order to cool the core.

Action 9. The assistant senior reactor operator looks toward HPCI cooling while the reactor operator attempts to recover the IC. HPCI also is inoperable in this sequence, leaving no operating high pressure coolant injection systems.

Figure 13-4 shows the ASRO noting the lack of HPCI flow.

176   Human Reliability Analysis

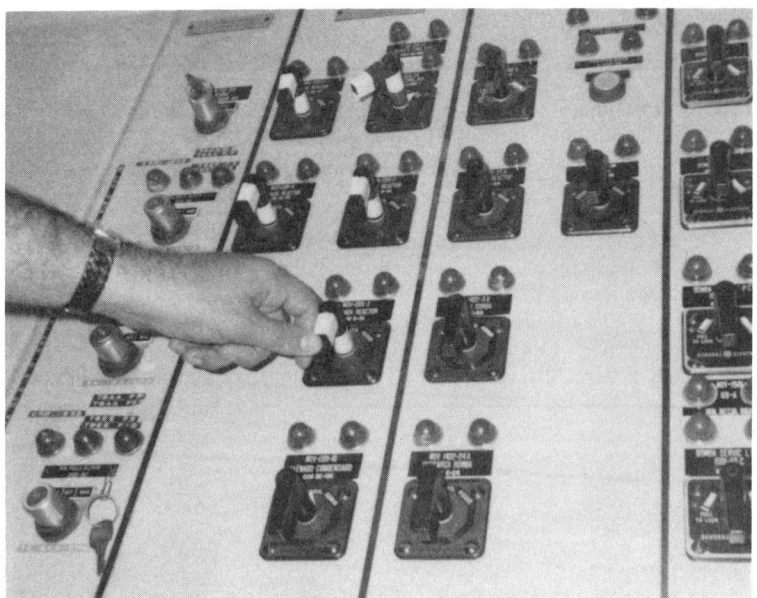

Figure 13-3. Operator attempts to obtain IC flow.

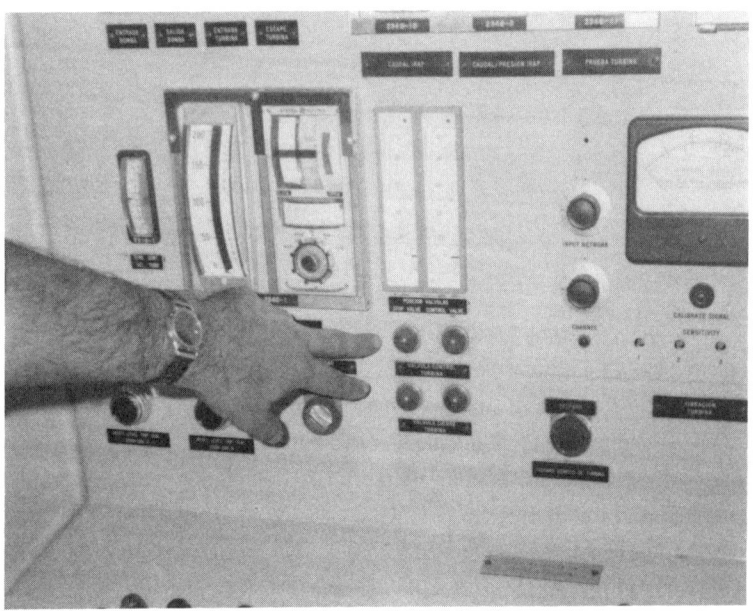

Figure 13-4. Lack of HPCI flow is noted.

Garoña Example 177

<u>Action 10</u>. Without some high pressure injection, reactor vessel level will continue to drop until near the top of active fuel (TAF). The POEs call for manual depressurization of the reactor. First, however, the ASRO will verify the operability of either LPCI, the preferred system, or core spray, in order to inject water at low pressure.

Figure 13-5 shows the ASRO attempting to start the LPCI pumps. Shortly after this action, the sequence as assessed is over with complete failure of high and low pressure injection.

In summary, the photographic survey allows a record of the major activities in a sequence, their location in the control room, and the instrumentation and controls that are important to achieving the actions. It also shows the order of priorities of using available resources and defines the recovery strategies and decisions. As such, a photo survey is probably second only to simulator exercises in obtaining insightful input from plant operators.

Figure 13-5. Operator attempts to use low pressure systems.

Quantification

There are three types of human failure events that are of most concern relative to off-normal recovery activities. These are:

1. Unrecovered slips - the crew has decided on the course of action and has chosen or developed the mitigation procedure and just fails to implement the decision.

2. Commissions - the crew misdiagnoses the situation and acts properly according to this incorrect diagnosis.

3. Time resource failures - the crew, for unspecifiable reasons, does not accomplish the chosen (and correct) activity in time.

Slips. The crew structure at Garoña is such that a turbine operator controls the feedwater and condensate systems, a reactor operator attends the control rods and monitors reactor level and pressure, an assistant senior reactor operator monitors and controls the safety systems, and the senior reactor operator monitors and coordinates the activities and makes the major decisions. This is a highly redundant structure that, using NUREG/CR-1278 technology, means that the probability of recovery of a slip is on the order of 0.99 to 0.9999. Further, any slip that has a significant consequence will result in an entry condition to the EPG/POE system and will call immediate attention to the problem. The net is that there seems no reason to include unrecovered slips as probable human failure events.

Commissions. The effect of a commission, i.e., action based on misdiagnosis, could be far reaching. But again, the EPG/POE system and the crew structure is optimized to all but eliminate these actions going unmitigated. Commissions, thus, were not considered in the Garoña HRA.

Time Failures. Simulator data and limited field data indicate that a well specified recovery action set takes time. The distribution of response times over all "possible" crews yields a time reliability correlation between the time available to accomplish the activity and the probability of doing so. The probabilistically complementary curve gives failure probability versus time.

Thus, events that are time dependent can be characterized as follows:

1. Is the action called out explicitly as a response to one of the critical entry symptoms in the POEs? Such an action is said to be "rule-based" rather than dependent on "general diagnosis".

2. Is there any conflict that would invoke hesitation on the operators in deciding or implementing the action? Conflict will increase the uncertainty associated with the decision/action.

3. What is the available time? Available time is the time from the leading symptom to a point when the action would not accomplish the intended purpose, subtracting off the expected time that it would take to implement the action.

4. What are the other major influences on the likelihood of the action's success? Influences can include the quality of particular instruments, the effectiveness of training, etc.

Quantification is performed as described in Chapter 11.

In the O1 event above, in the first scenario, the available time is some 12 minutes. There is no conflict and the concern for the IC is a rule in the POE system. P1 is, thus, about 0.04; U is 0.4 and L is 0.004. With a success likelihood index of 0.75, "good" but not "excellent", P2 is about 0.01. Thus, P3, the mean, is about 0.03. The key to that scenario is that O1 is independent of O2F and, thus, there would be another factor to multiply by the 0.03.

Conclusions

The Garoña plant is a well-run BWR, whose low level of risk is significantly affected by the human reliability of its operators. The plant operators are experienced, often with much longer job term than of their counterparts in the US. In fact, there are operators that have been present during the historical occurrence of all of the dominant risk sequence initiators with the exception of ATWS. The BWROG EPG system has been effectively implemented into the Garoña emergency recovery preparedness and training and blends well with this experience base.

Except for ATWS, there are no single human recovery actions that would have to be performed in times shorter than an hour. The EPGs and plant resources typically guarantee multiple opportunities in cases in which the available time is less than an hour. There were no observed cases of conflict in the sequences reviewed, except for the initiation of standby liquid control system following a partial ATWS. Thus, this BWR seems to contrast with several PWRs in the US in that there is no median term, i.e., 20-40 minute, high conflict action required of operators analogous to the feed and bleed option. This is due probably to the fact that reactor vessel level is directly affected by the loss of feedwater in a BWR, where there is a time lag in a PWR.

As a final note, there is no evidence, from the Garoña study, that the success of operator recovery in accident conditions is dependent on typical human factors concerns, except as they might contribute to the "burden" of the operator (Chapter 12). A plant interface, once learned and practiced on, is usually more than adequate, particularly when sequence timing is forgiving, to accommodate an event.

Chapter 14
FEED AND BLEED: A PWR EXAMPLE

The so-called "feed and bleed" option to cooling the core in a PWR was "discovered" as a result of the Three Mile Island accident. It is an option that arises when the normal core cooling, heat transfer path from the primary water, through the steam generators, to the secondary water side is interrupted on the secondary side. The initiating event typically includes a loss of main feedwater flow and is followed by the coincident inoperability of the emergency (auxiliary) feedwater system. In this case, primary water from the emergency core cooling system can be fed into the primary vessel and bled out one or more of the primary power/pilot-operated relief valves (PORVs). The heat of the water would then be transferred to the containment volume and removed by containment safeguards systems, bypassing the secondary side systems. This option substitutes secondary/primary cooling for primary/containment cooling and is clearly perceived by plant operators as a last course action.

Feed and bleed is an option to the restoration of some means of secondary cooling. There is typically enough time to attempt this preferable restoration effort. Even if successful, opting for feed and bleed assures a lengthy plant outage to clean up the containment building and affect repairs. In all PWRs with a feed and bleed option available, the actuation of the option is left as a manual action by operators onsite at the time.

NUREG/CR-1278 (Swain and Guttmann, 1983) includes an example analysis of the failure to use feed and bleed, i.e., high pressure injection primary cooling, to recover from a total loss of steam generator feedwater cooling (pp. 21-1 through 21-14). The qualitative analysis in NUREG/CR-1278 is clearly incorrect and the reasons that this is so are crucial to understanding the role and state-or-art of HRA. The NUREG analysis obtained five failure modes. The dominant failure mode was identified as the failure of the operators to notice that emergency (auxiliary) feedwater pumps fail to start following the loss of main feedwater. However, there is no reason to believe that operators could fail to detect the loss of **all** feedwater to the steam generators in a plant as fully instrumented and alarmed as a nuclear power plant.

In reality, the human reliability issue relates to the decision to give up trying to restore secondary cooling (at least as the first option) and to use feed and bleed. This decision presents the plant operators with a decisional burden due to conflict: using feed and bleed will guarantee a long, costly shutdown and bring NRC personnel into the event followup. So the question is: how long will the crew delay the unpreferred option in trying to obtain the preferred option?

182 Human Reliability Analysis

There are also plant-specific perturbations on this dilemma. Some plants, because of their design, may have marginal feed or bleed capability, and plant operators may be unlikely to resort to feed and bleed even in a situation in which it may be effective. Davis-Besse was such a plant when a loss of all feedwater event occurred in June 1985. Figure 14-1 shows a timeline of the loss of feedwater incident (USNRC, 1985). [These are the events that were abstracted in the example in Chapter 12.] In this event, the criterion for going to feed and bleed appears to have been ignored at about the 13 min point because the shift supervisor was attempting to start an alternative motor-driven feed water pump (which was started at 16 min). [At this point, the operators were "stretching procedures", i.e., interpreting the intent of the procedure in the light of the real event rather than simply following the words of the procedure without thinking.] Waiting for this pump proved to be successful, since the maximum RCS temperature obtained was within tolerance and occurred only after the start of the pump. The steam generated from starting this pump allowed the auxiliary feedwater pumps to be restarted and the incident was terminated safely. It should be noted that the operations superintendent had instructed the assistant shift supervisor to go to feed and bleed within one minute had the auxiliary feedwater not been restored, which introduced command and control burden. Just how long the crew might have waited had this activity not proceeded as smoothly is a matter of speculation.

In other plants, the secondary feedwater systems are so redundantly configured that a total, unrecoverable loss is considered incredible by operators and plant staff. The result is that, because of their belief structures, operators also could delay feed and bleed initiation too long. Still other plants have relatively low reliability EFW systems but more than adequate HPI and their crews may actually be "feed and bleed happy", i.e., they may resort to feed and bleed too quickly or in inappropriate situations. All of these perturbations have been noted in working with various plants' operations people.

Since the decisional elements of the feed and bleed option were omitted in the NUREG analysis and in fact cannot be explicitly accounted for according to the assumptions of THERP, a multiple-TRC approach is required. PRAs typically estimate that the time from the compelling signal for feed and bleed to significant core damage is on the order of 30 to 60 minutes. All PWRs that have feed and bleed as a viable option have written rules for its initiation. The cue consists of the loss of subcooling margin and initiation of a rise in core temperature, according to most feed and bleed rules. Thus, the rule-based TRCs of Figure 11-4 and Tables 10-8 through 10-9 may be used. These TRCs give failure probabilities of 0.01 to 0.002, for 30 and 60 min, when hesitancy and a SLI of 0.5 are assumed and 0.0003 to 0.000006 without hesitancy present.

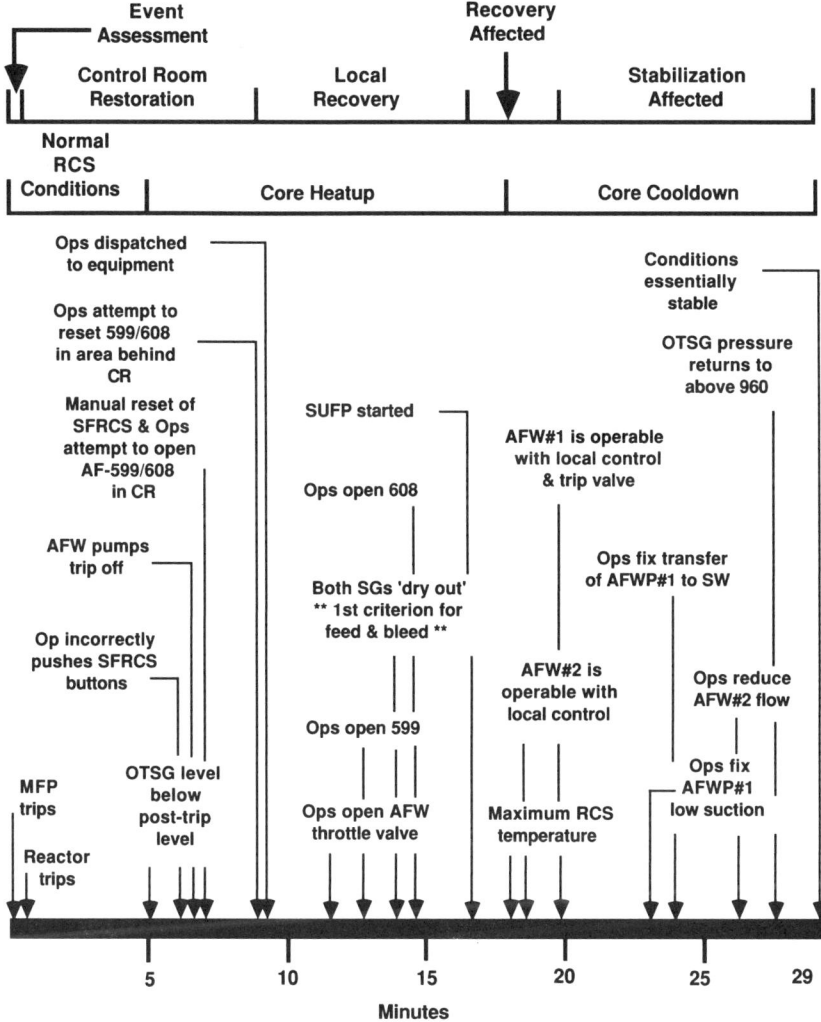

Figure 14-1. The June 1985 Loss of Feedwater at Davis-Besse.

It is assumed that in feed and bleed (F&B), the significant influences other than time and decision conflict are identified in interviews with operations personnel to be:

1. the training awareness of the action (denoted as training),
2. the clarity of the procedural guidance relative to the act (procedure),
3. the lack of a direct indication of the need to act (I&C).

These are assessed according to the procedure listed in Table 9-6. The first two influences are of a good-to-bad kind and the third is bad-only (see the terminology in Chapter 11). Table 14-1 shows the results.

Table 14-1
A SLI Calculation

Influence	Rank	Normed-rank	Quality	Product
training	30	0.6	0.8	0.48
procedure	10	0.2	0.7	0.14
I&C	10	0.2	0.5	0.10
	50			0.72
				SLI = 0.7

It is assumed further that the influences denoted "I&C" and "procedure" are found to have least impact on the act while "training" has 3 times the impact of these two. The these influences are ranked accordingly and are normalized as in the second column of Table 14-1. Finally, it is assumed that the quality of each influence, as measured by 1 for the best as licensed and 0 for the worst as licensed, is as in column three. The dot product is the sum in the fourth column and yields a success likelihood index of 0.7.

Using Table 10-9, for 30 minutes, a SLI of 0.7 yields a mean failure probability of 0.006. For 60 min, the failure probability is 0.0009. The values that would be obtained assuming hesitancy is not a problem (Table 10-8) are 0.00006 for 30 min and 0.000001 for 60 min. Thus, the final point estimates, which can vary by two to three orders of magnitude, are determined only on resolution of the hesitancy issue.

A residual issue relative to the NUREG is what to do with the other failure modes assessed for failing in feed and bleed recovery. This analysis shows that these modes are not dominant. The presumed dominant mode—failure to verify EFW pump start—is assessed at a probability of 0.0016, which is remarkably high considering its credibility. Also, since the example gives no credit for the Shift Technical Advisor nor anyone in the Technical Support Center, and because the loss of subcooling margin, which is modeled in the TRC approach, is so intimately associated with loss of EFW, this mode can also be dropped from further consideration.

Chapter 15
CONCLUSIONS

The HRA approach presented in this book is admittedly speculative, yet no more so than techniques already widely used in HRA and PRA. The approach has the strength of carrying to its logical conclusions a program of developing human reliability technology, which began with the Sandia human reliability handbook, NUREG/CR-1278 (Swain and Guttmann, 1983). The result is a robust enough accounting of many, if not all, of the types of human interactions involved in accomplishing the mission of operating a process facility. Although this approach is still under development, it has been used successfully in eleven PRA human reliability analyses and is offered as an interim solution to real and important human reliability problems.

Several main conclusions can be drawn from the development of such an HRA approach. These are:

1. The principal source material of HRA technology is the continuing exploration into the theory of individual and group failures. This theory is a burgeoning area of scientific study and is slowly uncovering why we do what we do. Human behavior theory and HRA technology are inexorably linked.

2. The loose connection between human error and human failure must be recognized and accommodated in the assessment of high-risk, complex technologies. This connection has implications for operation, design, and assessment. This connection can be initially examined using PRA and other analytical tools currently available.

3. Simulators are currently the best available sources of surrogate data in highly redundant technologies whose accidents are rare events, such as nuclear facilities. Plants should purchase simulators both to solidify training and to create an industry-wide database helpful to every plant operator.

4. The analysis of human failure events into their composites of modes, mechanisms, and causal factors is mature enough to guide risk management activities and pursue error reduction programs in process facilities. The qualitative analysis of an HRA is a crucial step in the credibility and utility of these efforts.

However, it should be recognized that there are limitations to this and any HRA approach and that there are areas that need further investigation. These are discussed in the context of the conclusions to be derived from the HRA approach presented.

Limitations

It should be clearly understood that the study of human behavior will always be a controversial area. Twenty five hundred years of studying human nature in the western world has seen precious little progress in understanding why we do what we do. The classification schemes used to identify human malfunctions are little more than what could have been expected from the Aristotelean Greek tradition; many of the monumental problems impeding the development of artificial intelligence, decision analysis, and the automating of process control today were first discussed in the time of Plato (Dreyfus and Dreyfus, 1986).

One of the more troublesome aspects of HRA is the fit of human performance to probabilistic quantification. The mechanisms and processes that intertwine to become human behavior seem to be succumbing to human scrutiny, yet at some low level, a quantitative fit is tenuous at best. Clearly, from a perspective of *predictability*, human-induced, high-risk events will remain stochastic. Yet there is no evidence that the mechanisms and capabilities that underlie human performance are in any way random. This is no different from the situation that exists for hardware. There, each failure occurrence can be said to have a definable cause; however, a specific failure occurrence must be considered a random event. As with hardware, randomness can often be considered as but a guise for ignorance, whether that ignorance can be removed or not. Thus, the occurrences of failure of either hardware or people can be reduced, if not eliminated, by identifying and addressing their causes.

Even though the theoretical probabilistic tie is no worse for the human element than it is for hardware elements, the historical basis for quantification of this tie is weaker for the human performance problem. While there are sources of human error data as discussed in Chapter 5, and while the current state of affairs is far better than it was just a decade ago, there is still a dearth of data. This situation often forces the HRA practitioners to use judgment to extend the existing scant data set. The introduction of these judgmental extensions even when sytematically applied, and even when the reasons for the extension are clearly indicated (as recommended here), weakens the credibility or increases the uncertainty of the resulting estimates.

The other major limitation of the HRA approach presented here is in the relationship between individual and group failure phenomena. Most of the current error research is applicable to the individual only. Group dynamics (see Janis, 1982, for example) is not simply the sum of individual behaviors. The simulator data, so far, only deals with group indices and thus, the reliability measures in this book are generally crew measures. However, this renders the assessment of crew structure, command and control, and other group factors impossible. This status will remain until that time the social interactions achieve the theoretical status that individual interactions will achieve.

Future Directions

Despite the inherent limitations of HRA, it has led to an understanding of human performance which, in turn, has demonstrably led to the reduction of risk in systems operation. Several areas would seem most ripe for investigation that would enhance the HRA state-of-the-art.

Surprisingly, no study has been sponsored by the nuclear industry or its regulators that examines all major incidents in nuclear power plants for their insights or implications relative to the influence of human interaction. Individual events have been examined (Kemeny, 1979; Chexal and Wycoff, 1980; Woods, 1982; and USNRC, 1985) but no integrated effort has been made to see what people do in a nuclear accident and whether this matches what HRA suggests they do.

Similarly, a study of long-term, low probability events, which dominate maritime accidents and are prevalent in oil drilling accidents and are conceived as important in a nuclear setting as well, could provide information which could allow analysts to understand the role of plant recovery teams and various contingency resources, such as technical support centers and evacuation plans, that have evolved mostly in the post-TMI era. The role of contingency training, i.e., the information relative to plant system connectivity and interaction, that is usually an informal part of training, if at all, should be examined for its effectiveness in emergency preparedness and response. Hesitancy during decision making, due to goal conflict or other sources of burden, has been observed in real events (Woods, 1982 and USNRC, 1985) and simulated exercises. The systematic study of the phenomena of burden or hesitancy could go far toward providing insights which would allow analysts to better understand human reliability characteristics under unanticipated conditions.

It is expected that efforts will continue to develop taxonomies and techniques for use in HRA in the nuclear industry. In addition, HRA should and probably will infiltrate the chemical and petrochemical process industry, where risks are not hidden by rarity and the regular costs of error can exceed those found in the nuclear industry. One study effort that could have immediate payoff is the investigation of methods of integrating human interaction considerations into the hazard analysis and loss prevention techniques already used in the process industries. It would appear that qualitative HRA could be integrated easily with these qualitative approaches, including the classical human factors reviews and the newer error analysis techniques used in HRA.

The human element in risk from our technologies will always be significant and perhaps dominant. HRA can help understand and control this risk and the approach presented is an interim step along the way toward the development of a comprehensive human reliability technology.

Appendix

INTRODUCTION TO PROBABILISTIC RISK ASSESSMENT FOR NUCLEAR POWER GENERATING STATIONS

Since the WASH-1400 study (USNRC, 1975), risk assessment models and tools have evolved to a point where they can provide plant decision makers with an accessible tool for achieving safe, reliable operation of nuclear power stations. The technical approach described in this appendix is only one way to perform a PRA. However, this approach is the culmination of 14 years and more than 1.5 million person hours of engineering, methodology research, applications, and software development and has been applied in the performance of some twenty PRAs of nuclear facilities, including the milestone Oconee PRA (NSAC, 1984).

Varieties of PRA

Nuclear station PRAs come in different varieties depending on the risk criterion to be analyzed and the scope of the events to be included (USNRC, 1983). Briefly, there are three levels of PRA, each successive level encompassing the prior. These levels correspond to three barriers between safe reactor operation and undesired consequences—core protection, containment, and distance. The scope of events assessed are also partitioned into two groups.

1. <u>Level 1</u>. The risk criterion in a level 1 PRA is the estimated frequency of major damage to the reactor core. Its risk management goal is the prevention of TMI-like accidents, where at TMI the public was not jeopardized but the utility lost a functioning unit. The phenomena of core melt, containment phenomena leading to radioactive material release, and health effects are not considered in a level 1 PRA.

2. <u>Level 2</u>. The risk criterion in a level 2 PRA is the estimated magnitude of major releases of radioactive materials and the occurrence frequencies of these consequences following containment failure. Its risk management goal is the prevention of accidents in which the public can be jeopardized because of a breach in containment. The health effects of the postulated accidents are not considered in a level 2 PRA.

3. <u>Level 3</u>. The risk criterion in a level III PRA is the estimated magnitude of public health consequences, including postulated deaths, as well as economic losses, and the occurrence frequencies of these consequences. Its risk

management goal is the prevention of public health consequences. The total risk equation for a nuclear station is considered in a level 3 PRA.

4. External Events Analysis. The scope of events considered in a PRA of any level can exclude so-called external events, such as seismic events, internal or external floods, fires, hurricanes or other damaging winds, transportation of hazardous (possibly non-nuclear) materials, etc. These are single "shock" events that the environment can impose on the plant, such as a seismic event, or plant events that can propagate with major impact throughout the plant, such as fires.

The inclusion or not of external events means that there are six varieties of PRA, the three levels with and without external events. WASH-1400 was a level 3 PRA that included a scoping study of external events. Many subsequent PRAs have been of the lower level varieties or have excluded external events, either to limit cost or to focus on the TMI-like category of accidents. Since the techniques of HRA described in this book are applicable to any level of PRA and to external events as well, only the commonly opted level 1 PRA, without external events as well as internal events, will be described in detail.

The Level 1 PRA

A level 1 PRA must assess the risk to a nuclear station of major core damage from internally-induced accidents. To assess the core damage risk, a plant risk model must be developed with sufficient detail to allow for the perturbation of model parameters or the models themselves, while maintaining its simplicity. This is a necessary characteristic in order to exercise the models to evaluate, for instance, the impact of assumptions or proposed plant changes. Adequate consideration must also be given to all support systems without producing unmanageable models. However, a tradeoff must be made between the detail of the modeling and the amount of work necessary to solve the models. The following approach focuses on the inclusion of appropriate model detail, while improving the computational and data handling capability of PRA tools to allow for efficient processing of the risk models.

A level 1 PRA model contains elements and structures necessary to assess changes in core damage frequency due to:

1. Modifications to component and system design to the level of detail available in system models, including all relevant support systems.

2. Changes in operation and maintenance procedures.

3. Changes in NRC licensing requirements by evaluating specific effects of those license requirements on component operability, operator actions, and system configuration. The PRA model allows investigation of alternatives to new NRC requirements as well as identifying risk-effective ways to meet them.

4. Changes in actual component and systems availability as plant-specific operation data becomes available and is incorporated into the PRA model.

5. Changes in the technology associated with human reliability as they relate to human behavior, human reliability data, and models.

6. Changes in technical specifications.

7. Changes in safety margins identified by core and plant deterministic analysis. Changes in accident sequence definitions and success criteria can be accommodated.

Through feedback of information from the human reliability analysis and the systems analysis tasks, insights can be obtained to support the existing operator training programs. Risk insights and operator actions identified during the recovery phase of the analysis can be of benefit to operator training programs. Dominant accident sequences and possible recovery actions can be included in training scenarios and simulated by simulators.

The nature of the risk model (comprehensive, yet easy to solve) provides an efficient capability for identification of the risk impact of planned work to support prioritization. The risk reduction benefits of projects in safety-related areas that improve plant availability can also be readily identified through the risk model. Extendibility of the plant model to include level 2, 3, and external events analysis is designed into the risk analysis approach and software.

The following sections provide a description of the technical effort associated with a level 1 PRA. In addition to providing core damage frequency estimates for internal initiators and identifying dominant contributors to those estimates, this PRA approach provides the structure necessary to integrate the optional external events analysis.

Organization of a Level 1 PRA

The overall approach to PRA is best described as selectively comprehensive—all system and human interactions are explicitly modeled, but only to the level of detail required to usefully assess a chosen risk criterion, in this case, core damage frequency. The approach is top-down, in that the model is incrementally developed from an initial general, broad-sweep perspective, down to further detail, including additional support requirements, as it is required. The method does not require all interactions and events to be addressed initially.

As noted in Chapter 7, risk can be correlated to sequences of events that lead to off-normal conditions. The development of a level 1 risk model thus requires the following tasks:

1. Identification of events, called initiators, that could lead to an off-normal condition, which in turn, if other events occurred, could lead to core damage,

2. Determination of events that represent plant responses to those initiators in terms of functions required to provide protection against core damage and determination of the plant end state events that result if protection is not adequately provided,

3. Development of system models to describe how plant function failure events in terms of both equipment available at the plant and operator actions, and integration of these models into a computer representation of failure events for the core damage sequences,

4. Performance of a data analysis, resulting in estimates of component reliability parameters for the basic events, and the integration of these results into the computer models to permit quantification of each sequence,

5. Performance of a human reliability analysis, resulting in estimates of human reliability parameters for the human failure events, and the integration of these results into the computer models to permit quantification of each sequence,

6. Solution of the integrated models to characterize the dominant core damage event sequences and their contributors, to estimate the overall core damage frequency and its uncertainty, and to identify potential recovery actions, and

7. Recommendation of action items to reduce station risk and specification of the requirements of a station risk management program.

The first six tasks lead to the technical results that allow the assessment of core damage risk. The last task is not an explicit requirement of a level 1 PRA but allows the best use of the PRA. Figure A-1 shows how the technical tasks interrelate. The following sections provide an overview of the seven PRA tasks.

Appendix - Introduction to PRA   193

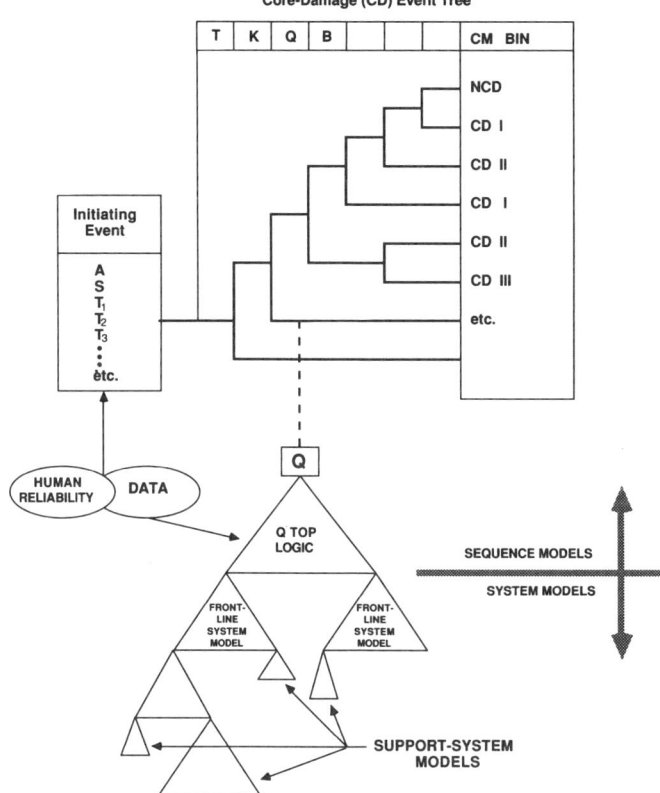

Figure A-1. The interrelationship of tasks in a level 1 PRA.

Initiating Events Analysis (Task 1)

The identification of a set of accident initiators is the first step in a PRA. Initiators are identified from a review of station's shutdown experience (if the plant has operated) and historic events for similar plants; those postulated in other PRAs; those considered in other risk methodologies; and those postulated based on a review of the station's design.

The functional plant response to an initiator is not necessarily unique. Some initiators produce the same demands on the plant and require the same systems to mitigate its effects. As a result, these initiators can be grouped together. Development of the complete set of initiators to be addressed and appropriate initiator groupings is typically a two-step process. A preliminary set of initiators is developed based on a review of historic events and initiators identified in other studies (see McClymont and Poehlman, 1982, for example). These initiators are used to define initiator groupings based on consistent plant response to initiators in a specific group. The system modeling task may also identify plant-specific initiators that are caused by failures within the systems modeled in detail but which are not identified in the initial review. These additional initiators are added to existing initiator groups, if applicable, or additional groupings are defined.

Accident Sequence Analysis (Task 2)

The characterization of plant responses to the initiators identified in Task 1 is the second step in a PRA. Plant response to initiators is described first using event trees. An event tree is constructed for each initiator group. The functions that can prevent the initiator from evolving into core damage are identified and arranged in a functionally logical, often chronological, order as event headers (represented by the capital letters in Figure A-1). Each event is a potential success or failure, represented as up or down branches on the tree. Some states preclude subsequent states, meaning that not all possible branches will be indicated.

An accident sequence then is a path through the branches of the event tree, consisting of an initiating event and a combination of various functional/system successes and failures that lead to an identifiable plant state. For example, loss of the secondary side cooling function would be modeled first rather than modeling the system failures of main feedwater and auxiliary feedwater as separate event tree branches. Core damage sequences are those paths of events that lead to core damage end states in the event trees.

This top-down approach allows the modeling to proceed in levels of increasing detail, from function to subfunction to system (see Figure A-2). Each function heading is logically decomposed into the individual system failure events, initiating events, and human failures that can cause the function to fail. These combinations are represented using small fault trees (Vesely, et al., 1981) and are referred to as "top logic". The top logic is then used to link the system fault tree models into an accurate and detailed representation of each core damage sequence. System fault tree models incorporate necessary support system models, as described in the next section. This approach provides the capability to directly model the details of systems interactions in a compact manner while being as comprehensive and as complete as possible.

An additional aspect of the event tree analysis is the inclusion of initiating events within the top logic. This allows one model to treat directly many potential initiating events which otherwise might be impractical to consider. Later, if it becomes desirable to include individual initiating events, and perhaps operator

Appendix - Introduction to PRA   195

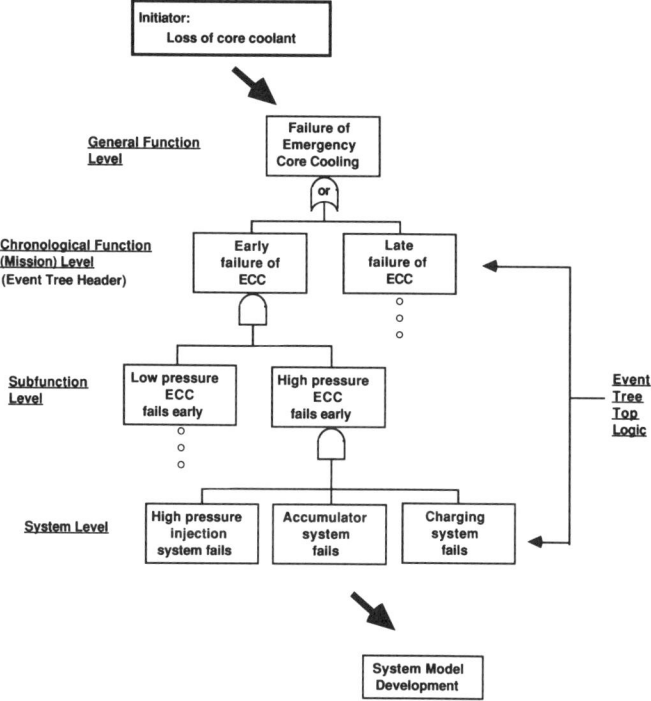

Figure A-2. The top-down approach to accident sequence identification.

actions, in the event tree headings, this can be done for informational purposes, while retaining the compact representation for ease of quantification.

It is recognized that the overall top logic can be equivalently modeled using a large event tree approach requiring many additional branches; however, many more accident sequences will be generated initially in the analysis. This functional event tree/top logic approach provides sufficient structure to define core damage sequences while making the review of the sequences easier and helping assure that all accident sequences have been identified.

Another effort in this task is the identification of appropriate core damage sequence end states and system success requirements. Thermal-hydraulic (T/H) modeling can be used to support elimination of unnecessary conservatism and increase confidence in the system success criteria and sequence and state definitions used in the project. Plant-specific T/H analyses strengthen the credibility of a PRA, offers opportunities for examining more realistic success criteria for identified plant-specific design vulnerabilities, and lays the groundwork for either addressing a variety of other issues such as containment integrity or performing a level 2 PRA.

The general approach is consistent with the *PRA Procedures Guide*, (USNRC, 1983). The approach consists of:

1. Determining the analyses required, the codes available and the potential benefits of plant specific analyses. Plant-specific analyses would only be performed in situations where substantial uncertainty or conservatisms are known to exist in current analyses, and only for significant accident sequences.

2. Collecting information for the required analyses and the code or codes selected to perform the analysis.

3. Developing models and support analyses for success criteria development and sequence and state definition.

Simplified thermal-hydraulic models can be used to support success criteria and sequence and state definitions. Selected benchmarks can be run against existing analyses to confirm the results of the simplified models. At this point, several accidents known to be of interest will be run and station personnel are trained in the use of the models. The intent of this effort is to provide station personnel with a tool to perform T/H analysis useful to gain confidence in the results of a level 1 PRA.

Sequence end states will be defined in terms of success and early and late core damage. Containment systems will be analyzed only for their impact on the likelihood of core damage. Limiting the scope of the analysis in this manner keeps the core damage models less complex. Further development of sequences and states to address overall plant damage states in preparation for proceeding to a level 2 or 3 analysis can be made by after the core damage models are completely solved. The existence of support system failures in the core damage cut sets (where applicable) simplifies determination of containment system response at this time. This determination can be performed with reasonable accuracy using the consequence parameter and range technique pioneered in the Oconee PRA (NSAC, 1984) without the need for additional thermal-hydraulic calculations due to the significant amount of published consequence information. If desired, however, other analysis methods can be called on to help verify and more precisely define the plant damage states.

Systems Modeling and Analysis (Task 3)

The top logic identifies the system failure events that are to be modeled in Task 3. Fault trees (Vesely, et al., 1981) are used to model the system failure events and are developed to the component level consistent with the level that failure data is available. The resulting models have a mathematical representation as Boolean logic equations for the top event. The Boolean equation is expressed in terms of "or" and "and" relationships of elemental events, the "leaves" of the fault tree. These elemental events are called "basic events" and sets of basic events that would cause the occurrence of the top event were they to occur are called "cutsets" (since together the events cut one or more success paths). This list of cutset descriptions of the system top events can be quantified by providing estimates of the failure rates and other reliability parameters, such as demand failure probabilities and unavailabilities, of the basic events in the models, typically equipment failures or human failures. The fault trees are organized along a train basis to allow independent groups of hardware failures to be grouped together into modules. These modules generally include all component failures associated with an equipment train. Figure A-3 is an example of the modularization of a simple system's fault tree. System dependencies associated with test and maintenance, common cause failure, and support systems will also be explicitly modeled in the fault trees.

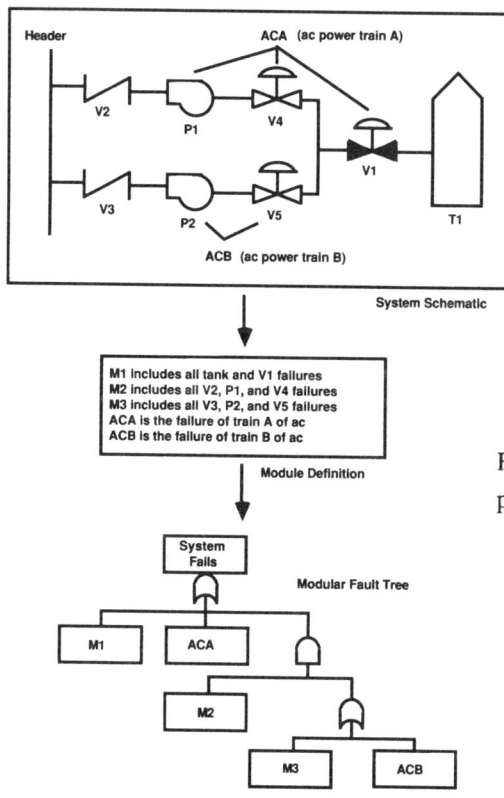

Figure A-3. The modularization process in systems modeling.

It is important to note that failures in support systems are directly modeled by linking support system fault trees to the front-line system trees. For example, if a valve requires electric power from a particular bus to operate, then failure to provide power to the valve is accounted for by a transfer to the portion of the electric power fault tree which models the bus failure. With this approach, failures of component in support systems are considered in the same way as failures of components in other systems. The impact of all component failures is embedded within the system and integrated into the plant models. The level of detail modeled can be revised during model development or at a later time to address particular concerns, such as the impact of instrument air failures on a specific system or detailed interactions in electric power systems. This approach greatly facilitates model modification to consider potential improvements, since the event tree/top logic structure seldom has to be modified.

This modularized approach has been extremely effective in allowing necessary dependencies to be rigorously modeled while reducing the computational requirements for solution of the accident sequences, which is an important feature since the model will be exercised many times in support of investigative activities after the PRA is completed.

Development of system models also provides information about potential initiators resulting from failures within systems. Examples of such events are cooling water train failures and electric bus failures which can result in a plant trip and degradation of systems required to mitigate the initiator. Information concerning these events are fed back to the initiating events and sequence modeling task.

Data Analysis (Task 4)

The quantitative risk criterion, core damage frequency, requires the use of data for estimating the basic failure parameters for events in the risk sequence models. If a plant has sufficient operating experience to allow the quantification of many basic events, then plant-specific information can be used almost exclusively. However quantification can also be provided by the means of a surrogate database constructed from the best available generic data sources. Even if a plant-specific database can be developed, a surrogate generic database will be needed to quantify those events for which there is insufficient operating experience. This database also provides a basis for comparison of the plant's experience against industry-wide experience.

The database includes a detailed description of the component type and failure mode for each of the generic component failure basic events developed by the system analysts. Along with each component type and failure mode combination, it will provide the associated mean, median, 5th and 95th percentile values, and the corresponding error factors. In addition the appropriate distribution parameters will be provided to allow for future updates by combining them later on with operating experience as it accrues. In a similar fashion generic maintenance unavailability data and initiator frequency data is provided where required for sequence quantification.

A generic database has the following disadvantages:

1. Generic data uncertainty bounds are often not good enough to perform risk applications such as technical specification review.

2. Regulatory bodies may not accept application for relief from regulations without the use of plant-specific operating information.

3. Generic data does not allow for the comparison of plant performance to generic performance and, perhaps more importantly, it does not allow comparison between systems or the root cause evaluations which are so essential for proper problem identification for corrective action.

When possible, a plant-specific data analysis is performed using the following tasks:

1. Data Identification and Collection. This work would involve the identification of data sources, including data from Licensee Event Reports (LERs) and the Nuclear Plant Reliability Data (NPRDs) system.

2. Establish Study Data "Windows". The initial work in the site specific data analysis is to establish the duration of the plant operating history which will be considered in the analysis.

3. The Data Encoding Process. The encoding process provides the foundation for the failure measure estimates. Failure categorization includes failure severity, failure mode, the proximate failure cause, the component type, the work order number or other record number, and the date for traceability back to the original records. The failure definitions and severity categories are established according to a failure categorization template that was developed as part of the In-Plant Reliability Data System Project (Drago, et al., 1982).

4. Component Exposure Calculation and Determination. The demand-related exposure is the historical number of demands experienced by the component population, demands due to periodic tests, automatic and manual initiation, failure related maintenance, and interfacing maintenance. The time-related

exposure corresponds to the historical time during the study data window during which the component population was operating.

5. Error Bound Determination. Calculating the error bounds is performed using the Chi-square distribution for the time-related events, and the F–distribution for demand related events when D - N <100 (where D = number of demands and N = number of failures). If D - N >100, then the Chi-squared distribution is used. The error bounds are then expressed as error factors, which are the ratio between the 95th percentile and the median.

6. Maintenance Unavailabilities. A structured interview technique is used to determine the contribution to a component's unavailability due to maintenance (Fragola and Collins, 1985). The approach uses the operational experience of plant personnel to establish the expected outage time distribution per occurrence for each equipment or train by systematically interviewing individuals and accounting for biases in their responses. This outage time data is then combined with the occurrence frequency data obtained from the workorders and other sources to provide an estimate of maintenance unavailability.

7. Combination of Generic and Site-Specific Data. Failure rates, failure probabilities and unavailability parameters will be derived using both the surrogate generic database and the plant-specific database. An engineering style of data combination is performed:

   a. If there is statistically sufficient data in the plant-specific sample and resulting estimate is within the generically estimated uncertainty range, then the plant-specific estimate is used.

   b. If the plant-specific estimate is not statistically sufficient then the plant-specific data is used to perform a Bayesian update of the surrogate generic data using the conjugate prior approach.

   c. If plant-specific data is unavailable or economically unattainable (often the case with instrumentation), then the surrogate generic data is used.

Human Reliability Analysis (Task 5)

The human reliability analysis is performed as described in Chapters 9 and 11.

Appendix - Introduction to PRA 201

Integral Plant Model and Quantification (Task 6)

The overall core damage risk model is developed by linking fault trees through the top logic defined during accident sequence development in Task 2. Detailed representations of accident sequence cutsets are called cutsets, which are obtained by computer solution of individual accident sequences.

Once the accident sequence cutsets are obtained, they will be reviewed to determine their validity and to identify critical parameters, such as accident timing, to be used to determine mitigating systems and actions available to the operator to effect recovery. Review of the cutsets for validity and removal of inconsistent cutsets, once an arduous task, is aided by software provided as a part of the PRA effort, which greatly simplifies the review process. Applicable recovery events are added to each dominant cutset and the cutset is requantified. This analysis ensures a realistic representation of the actual accident sequences.

The risk model will be developed and maintained using an integrated data base software. All system failure models and many of the accident sequences can be solved using this PC-based software. For sequences too large for microcomputer solution, mainframe codes are used for accident sequence quantification.

Once the model has been evaluated, the results of the evaluation are automatically loaded into a scoping model. The risk model is stored in terms of accident sequence cutsets made up of initiating events, hardware modules, components, human failures, and recovery failures. The hardware modules are also be stored within the system to allow for component level insights and sensitivities. The system can be used to select and rank accident sequence cutsets, calculate system, module, human failure, and component importance measures, as well as provide the capability to update model parameters and recalculate core damage frequency in a matter of minutes.

Risk Management (Task 7)

The risk management task serves as a focus for the major documentation efforts that are a part of a level 1 PRA:

1. The PRA final report. The final report and documentation are important to assuring traceability, as well as ease in updating and applying, the PRA.

2. The Risk Management Plan. The objectives of a risk management program is to provide a system for understanding, reducing and controlling the risks associated with the operation of the plant. A risk management program

provides a focused and comprehensive approach for cost-effectively achieving goals for plant risk, and requires a comprehensive plant-specific risk model that allows risks to be quantified, contributors to be pinpointed, and cost-effective changes for controlling and, if appropriate, reducing risk to be identified.

3. Recommendations for Risk Reduction. A third report in this task makes use of the results of Task 6 to identify potential areas for risk reduction. The report concentrates on practical measures, and addresses both procedural and hardware items. The report identifies potential changes and discusses advantages and disadvantages associated with each change (including uncertainty in the risk benefit estimate). Specific recommendations concerning necessary changes are avoided unless the expected impact is substantial and the change practical.

ACRONYMS

The following acronyms are used in the book.

ADS	automatic depresurrization system
AFW	auxiliary feedwater system
AIPA	accident initiator progression analysis
ASRO	assistant senior reactor operator
ATWS	anticipated transient without scram
BWR	boiling water reactor
BWROG	boiling water reactor owner's group
CCDF	complementary cumulative distribution function
CDF	cumulative distribution function
CSS	core spray system
DG	deisel generator
DSS	dedicated shutdown system
ECCS	emergency core cooling system
EdF	Electricité de France
EFW	emergency feedwater (system)
EOP	emergency operating procedure
EP	emergency procedure
EPRI	Electric Power Research Institute
ERG	emergency response guideline
HAZOP	hazards and operations analysis
HCR	human cognitive reliability (model)
HED	human engineering deficiency
HPCI	high pressure coolant injection (system)
HRA	human reliability analysis
IC	isolation condenser
IEEE	Institute of Electrical and Electronics Engineers
INPO	Institute of Nuclear Power Operation
IPRD	In-Plant Reliability Data (program)
IREP	Interim Reliability Evaluation Program
LER	licensee event report
LILCO	Long Island Lighting Co.
LOCA	loss of coolant accident
LOSP	loss of station (ac) power
LPCI	low pressure coolant injection
LWR	light water reactor

MORT	management oversight and risk tree (method)
MSIV	main steam isolation valve
NI	nuclear instrumentation
NPRDS	national plant reliability data system
NREP	national reliability evaluation program
NSAC	Nuclear Safety Analysis Center
OAET	operator action event tree
OAT	operator action tree
ORCA	operator reliability calculation and assessment
ORNL	Oak Ridge National Laboratory
POE	[translated from Spanish] emergency operating procedure
PORV	power- or pilot-operated relief valve
PRA	probabilistic risk assessment
PSF	performance shaping factor
PWR	pressurized water reactor
RCIC	reactor core isolation cooling (system)
RCS	reactor coolant system
RHR	residual heat removal
RPV	reactor pressure vessel
RSCS	reactor shutdown coolant system
Rx	reactor
SAT	station auxiliary transformer
SGTL	steam generator tube leak
SGTR	steam generator tube rupture
SHARP	systematic human action reliability procedure
SI	safety injection
SLCS	standby liquid control system
SLIM	success likelihood index methodology
SP	suppression pool
SRO	senior reactor operator
TAF	top of active fuel
TDP	turbine-driven pump
T/H	thermal hydraulic
THERP	technique for human error rate prediction
TMI	Three Mile Island (Unit 2)
TRC	time reliability correlation
TVA	Tennessee Valley Authority
USDoD	US Department of the Defense
USNRC	US Nuclear Regulatory Commission
VP	vent and purge (system)

BIBLIOGRAPHY

J. F. Ablitt, *A Quantitative Approach to the Evaluation of Safety Function of Operators in Nuclear Reactors*, AHSB(S) R-160, UKEA Health and Safety Branch, England, 1969.

J. R. Adams, "Issues in Human Reliability", *Human Factors*, vol. 24, no. 1, February 1982, pp. 1-10.

H. Anderson and B. Liwång, *Human Errors in Test and Maintenance of Nuclear Power Plants*, NKA/LIT(85)2, Nordic Liason Committee for Atomic Energy, August 1985.

W. B. Askren (ed), *Symposium on Reliability of Human Performance in Work*, AMRL-TR-67-88, Aerospace Medical Research Laboratories, Wright-Patterson Air Force Base, Ohio, May 1967.

L. J. Bain, *Statistical Analysis of Reliability and Life-Testing*. NY: Marcel Dekker, Inc., 1978.

L. Bainbridge, "Forgotten Alternatives in Skill and Work-load", *Ergonomics*, vol.21, no. 3, 1978, pp. 169-185.

R. E. Barlow and F. Prochan, *Mathematical Theory of Reliability*. New York: John Wiley & Sons, 1965.

I. Bazovsky, *Reliability: Theory and Practice*. New Jersey: Prentice-Hall, 1961.

A. N. Beare, et al., *Criteria for Safety-Related NPP Operator Actions: Initial BWR Simulator Exercises*, NUREG/CR-2535, Oak Ridge National Laboratory, November 1982.

C. R. Bell, "Psychological Aspects of Probability and Uncertainty", in C. R. Bell (ed.), *Uncertain Outcomes*, London: MTP Press, Ltd., 1979, pp. 5–21.

C. Berliner, et al., "Behaviors, Measures and Instruments for Performance Evaluation in Simulated Environments", paper presented at a Symposium and Workshop on Quantification of Human Performance, Albuquerque, NM, 1964.

Z. W. Birnbaum, D. J. Esary, and S. C. Saunders, "Multi-Component Systems and Structures and Their Reliability", *Technometrics*, vol. 3, no. 1, February 1961.

R. L. Black, "Observation Summary for B&W Operator Burden Task", letter to J.R.Fragola from Babcock & Wilcox Owners Group Engineering Services, February 25, 1987.

M. Bongard, *Pattern Recognition*. New York: Spartan Books, 1970.

T. F. Bott, E. Kozinsky, C. Crowe, P. M. Haas, *Criteria for Safety-Related NPP Operator Actions: Initial PWR Simulator Exercises*, NUREG/CR-1908, Oak Ridge National Laboratory, September 1981.

D. E. Broadbent, *Perception and Communication*. NY: Pergamon Press, 1958.

F. P. Brooks, "No Silver Bullet: Essence and Accidents of Software Engineering", *Computer*, vol. 20, no. 4, April 1987, pp. 10-19.

W. Brown and R. Wyrick (eds.), *Analysis of Steam Generator Tube Rupture Events at Oconee and Ginna*, INPO 82-030, Institute for Nuclear Power Operations, 1982.

R. L. Brune, M. Weinstein, and M. E. Fitzwater, *Peer Review Study of the Draft Handbook for Human Reliability Analysis with Emphasis on Nuclear Power Plantt Applications, NUREG/CR-1278*, Human Performance Technologies, SAND 82-7056, Sandia National Laboratories, January 1983.

S. K. Card, T. P. Moran, and A. Newell, *The Psychology of Human-Computer Interaction*. NJ: L. Erlbaum Associates, 1983.

A. B. Cherns in *Society: Problems and Methods of Study*, A. T. Welford, ed., London, 1962, pg. 162.

B. Chexal and H. Wycoff, *Instrument and Control Bus Power Loss at Rancho Seco on March 20, 1978*, NSAC-13, Electric Power Research Institute, November 1980.

M. K. Comer, D. A. Seaver, W. G. Stillwell, and C. D. Gaddy, *Generating Human Reliability Estimates Using Expert Judgment*, NUREG/CR-3688, for the USNRC, Washington, DC, November 1984.

M.K. Comer, E.J. Kozinsky, J.S. Eckel, and D.P. Miller, *Human Reliability Data Bank for Nuclear Power Plant Operations, Volume 2: A Data Bank Concept and System Description*, NUREG/CR-2744, February 1983.

M.K. Comer, M.D. Donovan, and C.D. Gaddy, *Human Reliablity Data Bank Evaluation Results*, NUREG/CR-4009, January 1985.

M.D. Comer and M.D. Donovan, *Specification of a Human Reliability Data Bank for Conducting HRA Segments of PRAs for Nuclear Power Plants*, NUREG/CR-4010, January 1985.

Conference Record for the 1981 IEEE Standards Workshop on Human Factors and Nuclear Safety, Institute of Electrical and Electronics Engineers, 1982.

D. C. Dennett, *Brainstorms*. Vermont: Bradford Books, 1978.

D. C. Dennett, *Elbow Room: The Varieties of Free Will Worth Wanting*. Masssachusetts: The MIT Press, 1986.

B. S. Dhillon, *Human Reliability with Human Factors*. NY: Pergamon Press, 1986.

J. P. Drago, R. J. Borkowski, J. R. Fragola, J. W. Johnson, *The In-Plant Reliability Data Base for Nuclear Plant Components: Interim Data Report—the Pump Component*, NUREG/CR-2886, Oak Ridge National Laboratory, December 1982.

H. L. Dreyfus and S. E. Dreyfus, *Mind Over Machine*, Free Press, 1986.

D. E. Embrey, *The Use of Performance Shaping Factors and Quantified Expert Judgment in the Evaluation of Human Reliability: An Initial Appraisal*, NUREG/CR-2986, Brookhaven National Laboratory, 1983.

D. E. Embrey, "SHERPA: A Systematic Human Error Reduction and Prediction Approach", *Proceedings of the International Topical Meeting on Advances in Human Factors in Nuclear Power Systems*, American Nuclear Society, April 21-21, 1986.

Electric Power Research Institute, *GO Methodology*, NP-3123-CCM, Palo Alto, California, June 1983.

K. N. Fleming, F. A. Silady, and G. W. Hannaman, "Treatment of Operator Actions in the HTGR Risk Assessment", *Transactions of the ANS 1979 Winter Meeting*, American Nuclear Society, November 1979.

J. R. Fragola, "Comment of Editorial: 'O Data, Data! Wherefore Art Thou Data?'", *IEEE Transactions on Reliability*, vol. R-32, no. 1, April 1983, pg. 2.

J. R. Fragola, *Identification, Categorization, and Analysis of Human Error Data Sources for Nuclear Power Plant Probabilistic Risk Assessment Applications*, Science Applications International Corporation, April 17, 1984.

J. R. Fragola and N. E. Chang, *Systematic Approach to Human Error Categorization for Procedural Behavior in Nuclear Power Plant Applications*, SAND83, Sandia National Laboratories, March 1983.

J. R. Fragola and E. P. Collins, "Interview Process for Obtaining Component Unavailability Data from Plant Personnel Experience", *International ANS/ENS Topical Meeting on Operability of Nuclear Power Systems in Normal and Adverse Environments*, Albuquerque, New Mexico, October 1986.

V. E. Frankl, *The Will to Meaning*. New York: New American Library, 1969.

R. R. Fullwood and A. A. Husseiny, "Human Performance and Response Time", *Transactions of the ANS 1979 Winter Meeting*, American Nuclear Society, November 1979.

C. Furst, *Origins of the Mind*. New Jersey: Prentice-Hall, Inc., 1979.

J. Galtung, "The Theory of Conflict and the Concept of Probability", in C. R. Bell (ed.), *Uncertain Outcomes*, London: MTP Press, Ltd., 1979, pp. 185-197.

M. S. Gazzaniga, *The Social Brain*. New York: Basic Books, 1985.

B. V. Gedenko, et al., *Mathematical Methods of Reliability Theory (English Translation of 1965 Publication)*. New York: Academic Press, 1969.

L. P. Goodstein, H. B. Anderson, S. E. Olsen (eds.), *Mental Models, Tasks, and Errors*. London: Taylor & Francis, expected in early 1988.

A. E. Greene, *Safety Assessment of Automatic and Manual Protection Systems in Reactors*, AHSB(S) R-172, UKEA Health and Safety Branch, England 1969.

D.I. Gertman et al, *Nuclear Computerized Library for Assessing Reactor Reliability (NUCLARR), Volume 1: Summary Description*, (Draft Rev. 1) NUREG/CR-4639, August 1987.

R. E. Hall, J. R. Fragola, and J. Wreathall, *Post Event Human Decision Errors: Operator Action Trees / Time Reliability Correlation*, NUREG/CR-3010, Brookhaven National Laboratory, November 1982.

G. W. Hannaman, A. J. Spurgin, and J. R. Fragola, *Systematic Human Action Reliability Procedure (SHARP)*, Interim Report, NP-3583, Electric Power Research Institute, June 1984.

G. W. Hannaman, A. J. Spurgin, and Y. D. Lukic, *Human Cognitive Reliability Model for PRA Analysis*, NUS-4531, NUS Corporation, December 1984.

R. L. Henneman and W. B. Rouse, "On Measuring the Complexity of Monitoring and Controlling of Large-Scale Systems", *IEEE Transactions on Systems, Man and Cybernetics*, vol. SMC-16, March/April 1986, pp. 193–207.

R. A. Hess, "A qualitative model of human interaction with complex dynamic systems", *IEEE Transactions on Systems, Man, and Cybernetics*, vol. SMC–17, no.1, Jan./Feb. 1987, pp. 33-51.

E. Hoffer, *The True Believer*. New York: Harper & Row, Publishers, 1951.

D. R. Hofstatder, *Metamagical Themas: Questing for the Essence of Mind and Pattern*. New York: Basic Books, 1979.

D. R. Hofstatder, *Gödel, Escher, Bach: An Eternal Golden Braid*. New York: Basic Books, 1985.

E. Hollnagel, O. M. Pedersen, and J. Rasmussen, *Notes on Human Performance Analysis*, RISØ-M-2285, RISØ National Laboratory, Denmark, June 1981.

R. A. Howard and J. E. Matheson, *Influence Diagrams*, SRI International Report, January 4, 1980.

Human Factors, Volume 6 (Summary of Papers Presented at a Symposium held by the Electronics Industries Association - Human Factors Subcommittee, Albuquerque, NM), December 1964.

Institute for Electrical and Electronics Engineers, *Guide for General Principles of Reliability Analysis of Nuclear Power Generating Station Protection System*, IEEE Std 352-1975 (ANSI N41.4-1976), April 14, 1975.

Institute for Electrical and Electronics Engineers, *Guide for General Principles of Human Action Reliability Analysis for Nuclear Power Generating Stations*, Project 1082, Nuclear Power Engineering Committee, Subcommittee 7, Human Factors and Control Facilities, Unpublished draft 4, March 1987.

Institute of Nuclear Power Operations, *NPRDS Reporting Procedures Manual*, INPO 84-011, Rev. 10, Atlanta, GA, April 1984.

I. L. Janis, *Victims of Groupthink*. Boston: Houghton Mifflin, 1982.

I. L. Janis and L. Mann, *Decision Making*, Free Press, New York, 1977.

J. P. Jenkins (ed), *Proceedings of the 1970 U. S. Navy Human Reliability Workshop*, July 22-23, 1970, Washington, DC, Navships 0967-412-4010.

R. S. Jensen, "Pilot Judgment: Training and Evaluation", *Human Factors*, vol. 24, no. 1, Feb. 1982, pp. 61-73.

G. Johannsen and W. B. Rouse, "Studies of Planning Behavior of Aircraft Pilots in Normal, Abnormal, and Emergency Situations", *IEEE Transactions on Systems, Man and Cybernetics*, vol. SMC-13, no. 3, May/June 1983, pp. 267-278.

W. G. Johnson, *MORT, Safety Assurance Systems*. NY: Marcel Dekker, Inc., 1980.

D. Kelley, *The Evidence of the Senses*. Louisiana: Louisiana State University Press, 1986.

Kemeny (President's Commission on the Accident at Three Mile Island), *Three Mile Island: A Report to the Commissioners and to the Public*, Washington DC, 1979.

G. J. Kolb, et al., *Interim Reliability Evaluation Program - Analysis of Arkansas Nuclear One, Unit One Nuclear Power Plant*, NUREG/CR-2787, Sandia National Laboratories, June 1982.

F. P. Lees, *Loss Prevention in the Process Industries*. London: Butterworth & Co., 1980.

E. J. Lerner, "Helping the pilot handle the supercockpit", *Aerospace America*, February 1987, pp. 28-31.

H. W. Lewis, R. J. Budnitz, H. J. C. Kouts, W. B. Lowenstein, W. D. Rowe, F. von Hippel, and F. Zachariasen, *Risk Assessment Review Group Report to the U. S. Nuclear Regulatory Commission*, NUREG/CR-0400, US NRC, Sept. 1978.

K. C. Lish, *Nuclear Power Plant Systems and Equipment*. NY: Industrial Press, Inc., 1972.

D. K. Lloyd and M. Lipow, *Reliability: Management, Methods and Mathematics*. New Jersey: Prentice-Hall, 1961.

Long Island Lighting Company with SAIC, *Probabilistic Risk Assessment - Shoreham Nuclear Power Station, Final Report*, LILCO, NY, April 1983.

Llory, Lemaitre, M. Griffon-Fouco, and T. Meslin, "Analysis of Operator's Behavior Under Accidental Transients", April 7, 1987 Meeting of Spanish Nuclear Society's Human Factors Division, Madrid, Spain.

W.J. Luckas et al, *A Human Reliability Analysis for the ATWS Accident Sequence with MSIV Closure at the Peach Bottom Atomic Power Station*, BNL Tech. Report A-3272, May 1986.

J. T. Macnamara, *Names for Things*. Massachusetts: The MIT Press, 1982.

H. F. Martz and M. C. Bryson, "A Statistical Model fo. Combining Biased Expert Opinion", *IEEE Transactions on Reliability*, vol. R-33, no 3, August 1984, pp. 227-232.

R. May, *The Courage to Create*. New York: W. W. Norton & Company, Inc., 1975.

R. May, *Freedom and Destiny*. New York: W. W. Norton & Company, Inc., 1981.

H. B. Maynard, G. J. Stegemerten, and J. L. Schwab, *Methods-Time Measurement*, McGraw-Hill Industrial Organization and Management Series, NY, 1948.

A. S. McClymont and B. W. Poehlman, *ATWS: A Reappraisal, Part 3: Frequency of Anticipated Transients*, NP02230, SAIC for Electric Power Research Institute, January 1982.

D. Meister, "Use of a Human Reliability Technique to Select Desirable Design Configurations", paper presented at the Eighth Reliability and Maintainability Conference, Denver, Colorado, July 1969, *1969 Annals of Assurance Sciences*.

G. A. Miller, "The magical number of seven plus or minus two: some limits on our capacity for processing information", *Psychological Review*, vol. 63, 1956, pp. 81-97.

J. W. Minarick and C. A. Kukielka, *Precursors to Potential Severe Core Damage Accidents: 1969-1979 A Status Report*, NUREG/CR-2497, SAIC for the USNRC, June 1982.

N. Moray (ed.), *Mental Workload: Its Theory and Measurement*. NY: Plenum Press, 1979.

S. J. Munger, R. W. Smith, and D. Payne, *An Index of Electronic Equipment Operability: Data Store*, AIR-C43-1/62-RP (1), American Institute for Research, Pittsburgh, PA, January 1962.

National Transportation Safety Board, *Eastern Airlines L-1011, Miami, Florida, December 29, 1972*, NTSB-AAR-73-14, Washington DC, June 1973.

U. Neisser, *Cognition and Reality*, San Francisco: Freeman Press, 1976.

D. Nielsen, "Use of Cause-Consequence Charts in Practical Systems Analysis" in *Reliability and Fault Tree Analysis Theoretical and Applied Aspects of System Reliability*, SIAM, Philadelphia, Pennsylvania, 1975, pp. 849-880.

D. A. Norman, "Position Paper on Human Error", *NATO Conference on Human Error*, Bellagio, Italy, September 1983.

R. Nozick, *Philosophical Explanations*. Massachusetts: Harvard University Press, 1981.

Nuclear Safety Analysis Center, *Analysis of Three Mile Island—Unit 2 Accident*, NSAC-1, Electric Power Research Institute, July 1979.

Nuclear Safety Analysis Center, *Oconee PRA - A Probabilistic Risk Assessment of Oconee Unit 3*, NSAC-60, vol. 1, Electric Power Research Institute, June 1984.

J.N. O'Brien and C.M. Spettell, *Uses of Human Reliability Analysis Probalistic Risk Assesment Results to Resolve Personnel Performace Issues that Could Affect Safety*, NUREG/CR-4103, October 1985.

I. A. Papazoglou, *Probabilistic Safety Analysis Procedures Guide*, NUREG/CR-2815, January 1984.

K. S. Park, *Human Reliability Analysis: Analysis, Prediction, and Prevention of Human Errors*. Netherlands: Elsevier Science Publishers, Co., 1987.

J. F. Parker and V. R. West (ed.), *Bioastronautics Data Book*, NASA, Washington, DC, SP-3006, 1973.

E. S. Pearson and H. O. Hartley, *Biometrika Tables for Statisticians, Volume I*. London: Cambridge University Press, 1954.

C. Perrow, *Normal Accidents*. NY: Basic Books, 1984.

L. D. Phillips, P. Humphreys, D. E. Embrey, and D. L. Selby, "A socio-technical approach to assessing human reliability", Appendix D of *A Pressurized Thermal Shock Evaluation of the Calvert Cliffs Unit 1 Nuclear Power Plant*, NUREG/CR-418, Oak Ridge National Laboratory, September 1985.

K. H. Pribram, "Problems concerning the structure of consciousness", in G. G. Globus, G. Maxwell, and I. Savodnik (eds.), *Consciousness and the Brain*, New York: Plenum Press, 1976.

Proceedings of the 1975 Annual Reliability and Maintainability Symposium, January 1975.

Reliability Analysis Center, *Component Reliability Data Books*, NPRD-2, RADC/RAC, Griffiss AFB, New York, 1984.

J. Rasmussen, "Outlines of a hybrid model of the process operator", in T. B. Sheridan and G. Johannsen (eds.), *Monitoring and Supervisory Control*, New York: John Wiley & Sons, 1976.

J. Rasmussen, "Skills, Rules, and Knowledge; Signals, Signs, and Symbols, and Other Distinctions in Human Performance Models", *IEEE Transactions on Systems, Man, and Cybernetics*, vol. SMC-13, no. 3., May/June 1983, pp. 257-266.

J. Rasmussen, A. Carnino, M. Griffon, G. Mancini, and P. Gagnolet, *Classification System for Reporting Events Involving Human Malfunctions*, RISØ-M-2240, RISØ National Laboratory, Denmark, March 1981.

J. Rasmussen, J. Leplat, and K. Duncan, *New Technology and Human Error*. NY: John Wiley & Sons, 1987.

J. T. Reason, "Absent-mindedness and Cognitive Control", in *Everyday Memory, Actions and Absent-Mindedness*, Academic Press, 1983, pp. 113–132.

J. T. Reason, "A Framework for Classifying Errors", in *New Technology and Human Error*, J. Rasmussen, et al., (eds.), NY: John Wiley & Sons, 1987.

J. T. Reason, "Generic Error-Modelling System (GEMS): A Cognitive Framework for Locating Common Human Error Forms", in *New Technology and Human Error*, J. Rasmussen, et al., (eds.), NY: John Wiley & Sons, 1987.

J. T. Reason and D. E. Embrey, *Human Factor Principles Relevant to Modelling of Human Errors in Abnormal Conditions of Nuclear and Major Hazardous Facilities*, Human Reliability Associates, 1 School House, Higher Lane, Dalton, Parbold, Lancashire, WN8 7RP, England, August 1985.

J. T. Reason and K. Mycielska, *Absent-Minded? The Psychology of Mental Lapses and Everyday Errors*, Prentice-Hall, 1982.

L. V. Rigby, "The Sandia Human Error Rate Bank (SHERB)", in Man–Machine Effectiveness Analysis, A Symposium of the Human Factors Society, Los Angeles Chapter, June 15, 1967, pp. 5-1 to 5-13.

M. Rogovin, director, *Three Mile Island, A Report to the Commissioners and to the Public*, NUREG/CR-1250, USNRC, January 1980.

S. P. R. Rose, *The Conscious Brain*. New York: Vintage Books, 1976.

W. B. Rouse, "Optimal allocation of system development resources to reduce and/or tolerate human error", *IEEE Transactions on Systems, Man and Cybernetics*, vol. SMC-15, no. 5, Sept./Oct. 1985, pp. 620-630.

G. H. Sandler, *System Reliability Engineering*. New Jersey: Prentice-Hall, 1963.

T. Schmall (ed), *Conference Record for 1979 IEEE Standards Workshop on Human Factors and Nuclear Safety*, Institute of Electrical and Electronics Engineers, 1980.

B. Shneiderman, *Software Psychology: Human Factors in Computer and Information Systems*. Massuchusetts: Winthrop Pulishers, Inc., 1980.

M. L. Shooman, *Probabilistic Reliability: An Engineering Approach*. New York: McGraw-Hill, 1969, new edition, NY: Kreiger, 1987.

K. N. Smith, *Elegy for a Soprano*. New York: Villard Books, 1985.

W. G. Stillwell, D. A. Seaver, and J. P. Swartz, *Expert Estimation of Human Error Probabilities in Nuclear Power Plant Operations: A Review of Probability and Scaling*, NUREG/CR-2255, USNRC, May 1982.

A. D. Swain, "Some Problems in the Measurement of Human Performance in Man-Machine Systems", *Human Factors*, vol. 6, 1964, pp. 687-700.

A. D. Swain, *Accident Sequence Evaluation Program Human Reliability Analysis Procedure*, NUREG/CR-4772, USNRC, Washington, DC, February 1987.

A. D. Swain and H. E. Guttmann, *Handbook of Human Reliability Analysis with Emphasis on Nuclear Power Applications*, NUREG/CR–1278, Sandia National Laboratories, August 1983.

D.A. Topmiller, J.S. Eckel, and E.J. Kozinsky, *Human Reliability Data Bank for Nuclear Power Plant Operations, Volume 1: A Review of Existing Human Reliability Data Banks*, NUREG/CR-2744, December 1982.

R. Trivers and H. P. Newton, "The Crash of Flight 90: Doomed by Self-Deception?", *Science Digest*, November 1982, p. 66.

A. Tversky and D. Khaneman, "Judgment under uncertainty: heuristics and biases", *Science*, vol. 185, September 1974, pp. 1124-1131.

U. S. Department of Defense, *Military Standard, Human Engineering Requirements for Military Systems, Equipment and Facilities*, MIL–STD–1472A, Washington, DC, May 15, 1970.

USDoD, *Human Reliability System Users' Manual*, USNAVSEA, Washington DC, December 1977.

USDoD, *Military Handbook - Reliability Prediction of Electronic Equipment*, MIL-HDBK-217D, Washington DC, January 1982.

USNRC (U. S. Nuclear Regulatory Commission), *Reactor Safety Study - An Assessment of Accident Risks in U. S. Commercial Nuclear Power Plants*, WASH-1400 (NUREG-75/014), Washington, DC, 1975.

USNRC, *Recommendations Related to Brown's Ferry Fire*, NUREG-0050, Washington DC, February 1976.

USNRC, *PRA Procedures Guide - A Guide to the Performance of Probabilistic Risk Assessment for Nuclear Power Plants*, NUREG/CR–2300, Washington DC, December 1982.

USNRC, *Licensee Event Report System, Description of System Guideline for Reporting*, NUREG-1022, Washington, DC, September 1983.

USNRC, *Loss of Main and Auxiliary Feedwater Event at the Davis-Besse Plant on June 9, 1985*, NUREG-1154, Washington, DC, July 1985.

H. P. Van Cott and R. G. Kincade, *Human Engineering Guide to Equipment Design*, U. S. Government Printing Office, Washington, DC, Revised Edition, 1972.

W. E. Vesely, F. F. Goldberg, N. H. Roberts, D. F. Haasl, *Fault Tree Handbook*, NUREG/CR-0492, US NRC, Washington, DC, 1981.

A. Villemeur, J. M. Moroni, F. mosneron-Dupin, and T. Meslin, "A Simulator-based Evaluation of Operators' Behavior by Electricité de France", *Proceedings of the International Topical Meeting on Advances in Human Factors in Nuclear Power Systems*, Knoxville, Tennessee, April 21-24, 1986, American Nuclear Society, pp. 374-379.

J. L. vonHerrmann, *Methods for Review and Evaluation of Emergency Procedure Guidelines vol. 1: Methodologies*, NUREG/CR-3177, Delian Corp. for USNRC, March 1983.

K.J. Voska and J.N. O'Brien, *Human Error Probability Estimation Using Licensee Event Reports*, NUREG/CR-3519, July 1984.

J. M. Waage, *Screening and Evaluation of 1979 Licensee Event Reports*, NSAC-9, Electric Power Research Institute, December 1980.

R. J. Waller, "Complexity and the boundaries of human policy making", *International Journal of General Systems*, vol. 9, 1982, pp. 1-11.

E. L. Weiner and R. E. Curry, "Flight Deck Automation: Promises and Perils", *Ergonomics*, vol. 23, 1980, pp. 995-1011.

A. T. Welford, *Fundamentals of Skill*. London: Methuen, 1968.

D. W. Whitehead, L. M. Weston, and N. L. Graves, *Risk Methods Integration and Evaluation Program Methods Development: A Data-Based Methodology for Including Recovery Actions in PRA*, vol. 1, draft, NUREG/CR-4834, Washington, DC, February 1987.

J. G. Wohl, "Maintainability Prediction Revised: Diagnostic Behavior, System Complexity, and Repair Time", *IEEE Transactions on Systems, Man and Cybernetics*, vol. SMC-12, no. 3, May/June 1982, pp. 241-250.

J. G. Wohl, "Cognitive capability versus system complexity in electronic maintenance", *IEEE Transactions on Systems, Man and Cybernetics*, vol. SMC-13, no. 4, 1982, pp. 624-626.

D. D. Woods, "Operator decision behavior during the steam generator tube rupture at the Ginna nuclear power station", in *Analysis of Steam Generator Tube Rupture Events at Oconee and Ginna*, INPO 82-030, W. Brown and R. Wyrick (eds.), Institute for Nuclear Power Operations, 1982.

D. D. Woods, "Graphic representations of complex worlds", *Proceedings of the 1986 IEEE International Conference on Systems, Man, and Cybernetics*, 86CH2364-8, Atlanta, Georgia, October 14-17, pp. 259-261.

D. D. Woods, E. M. Roth, and L. F. Hanes, *Models of Cognitive Behavior in Nuclear Power Plant Personnel*, NUREG/CR-4532, US Nuclear Regulatory Commission, Washington, DC, July 1986.

W. E. Woodson, *Human Factors Design Handbook*. McGraw-Hill, 1981.

GLOSSARY

The following terms are significant enough to the ideas in the book to warrant some explanation. The definitions provided are probably parochial, due to the authors' usage of the words, but they were chosen to be as close to "common-usage" as possible.

abnormal event see **off-normal event**

accident an occurrence that leaves a system damaged rather than just interrupting its performance

action the "output" from a person—speech, bodily movement, perception, or thought

annunciator a display that has an audible indication as well as visible

attention the human capacity or mechanisms of being aware or conscious of the environment, one's thoughts, or one's actions [consciousness is supposedly an enhanced, self-aware level of attention]

behavior the actions (or activity) of a person (or machine)

Boolean equation a symbolic equation with symbols standing for phrases connected by logical connectives such as "and", "or", "not", etc.

burden aspects of the environment or fundamental human mechanisms that put loads on human capacities for action or otherwise inhibit performance

cognition the capacity or mechanisms that lead to knowledge

cognitive science an interdisciplinary, scientific field that includes parts of psychology, philosophy, linguistics, neurobiology, artificial intelligence and other areas that study cognition

command (1) an electronic signal sent to a control device, or (2) an instruction sent to a co-worker [the intention is the same in either case: to change the state of a person or piece of equipment]

commission error an error that amounts to an unintended action, excluding inaction

common-cause failure a failure that leads to another failure through some system or environmental interaction

communication (1) the capacity or mechanisms of information transfer between or among people, or (2) the result of using this capacity

complexity a system property that arises when many components mutually interact so as to challenge the comprehension capabilities of people [complexity is not merely the fact of a large number of components]

confidence limit a number that delimits a parameter with some amount of confidence

consequence (1) the result of an event or action, or (2) the result as an event

control an electronic and/or mechanical device that can change the state of another device

control room a centralized location from which operators can perform the normal control of a system or plant

crew a group of people with a joint mission, e.g., the operating crew in a nuclear power plant

cutset a list of failure events (or their symbols) which, if occurring together, fail the system or lead to the sequence

cycle an operational loop of a device from one state through at least one other and back again

decision making (1) the activity of choosing one course of action among alternatives, or (2) the capacity or mechanisms that make decisions

demand (1) a signal or action that should change the state of a device, or (2) an opportunity to act, and thus, to fail

density a mathematical function that is integrable to unity, i.e., a **probability density function**

dependency a relationship between one event, e.g., the failure to notice an alarm, and another event, e.g., the failure of the alarm, in which the one causally succeeds from the other

dependent one event is dependent on another if a dependency exists between them

design the intended function, configuration, and operation of a device or system

diagnosis the capacity or mechanisms to understand what is perceived and realize the implications of a perceived situation

display a device that indicates a parameter of equipment's status to a person by some perceptual mechanism

distribution (1) the scattering of similar data in some coordinate system, or (2) a function that is the indefinite integral of a density, i.e., a cumulative distribution function

emergency situation or circumstances that are not as intended at the moment and which imply an incipient threat of damage, e.g., during an accident

environment everything "outside" a system being considered

error (1) any deviation from an intended or desired human performance, or (2) any deviation from a target (real or analytical)

estimate (1) human capacity to assess some attribute against a standard, or (2) an empirically produced number that may represent some ideal (reliability) parameter

event a point in time or some duration that marks a change in conditions from the past situation

event tree a graphical representation of the logic of the interactions of intermediate events between an initiator and its identified consequence(s)

external event a PRA term to indicate an environmental common-cause failure such as an earthquake, a flood, or a fire

failure any deviation from an intended or desired hardware, software, or system performance [note that an **error** is a failure of a person, whereas a **human failure** is a failure of hardware, software, or a system that is attributed to human action (which may or may not be in error)]

fault tree a graphical representation of the logic of the causes of failure of a specified (top) event

Gaussian distribution the standard normal distribution, centered about zero and symmetric, i.e., the so-called "bell-shaped" curve

hazard a feature of the environment that could be harmful or damaging to a system or person

hardware mechanical or electrical artifices

hesitancy a disinclination to act [can be brought on by burden, e.g., uncertainty]

human factors (1) any attribute of a situation or object that is due to the action(s) or attribute(s) of one (or more) person, or (2) the discipline that studies such factors

human performance the result of human behavior as measured against some goal or standard

human reliability the probability that the performance a person or group of people will be successful, i.e., acceptable against the standard or goal of the performance, over some mission, i.e., a duration or demand for the performance

incident an occurrence that interrupts the performance of a system rather than leaving a system damaged

influence a causal factor for a specific event

initiator (initiating event) the occurrence that starts an incident or accident, e.g., the first event in a sequence

intention (1) the capacity or mechanisms that allow the choice of a goal or purpose, or (2) the goal itself

220   Human Reliability Analysis

interaction the relationship between the behavior of two systems or components to produce a "combined" consequence that would not occur if only the behavior of the individual system or component occurred [a **human interaction** is an interaction in which at least one system or component is a person]

lapse an error in recall, e.g., of a step in a task or a name or a word [a type of slip]

maintenance (1) the upkeep of a system or device, (2) the staff of people charged with the upkeep

man-machine interface (user interface) the (abstract) boundary between people and the hardware or software they interact with

mechanism a function of the body, particularly the brain

mistake an error in establishing a course of action, e.g., an error in diagnosis, decision making, or planning

model (representation) a way of describing or conceptualizing a system and its interactions [as distinct from the scale models sometimes used in design engineering]

off-normal event [same as abnormal event] situation or circumstances that are not as intended at the moment, e.g., during an incident, but do not yet imply an incipient threat of damage

omission error an error that amounts to an unintended or unnoticed inaction

operating a state of a device in which it is dynamically performing one of its designed functions

operation (1) the putting of a device, a system, or several systems into one of their operating states, or (2) [plural] the staff of people charge with operating equipment

parameter a number [often produced by a signal, or other electronic measurement, or from a calculation]

perception the capacity or mechanisms that lead to recognizing sensory input

performance shaping factor an influence on performance

plan an overall strategy or sequence of procedures that are intended to achieve a goal

probability a number between 0 and 1, inclusively, that quantitatively ranks the likelihood or chance of the occurrence of a postulated event

procedure the formal realization of a task, e.g., verbal instructions or written procedure

random variability that cannot be predicted or its causes are unknown or its results have no discernible pattern

random variable a relationship between an unpredictable phenomenon and its probability of occurrence

recovery the accommodation of a failure or otherwise undesired performance in hardware or software by restoring the failed hardware or software or by finding an alternative to achieving the function of the hardware or software [error recovery is a recovery from one's own or another's error]

reliability the probability that the performance of some hardware, software, people, or their combination will be successful, i.e., acceptable against the standard or goal of the performance, over some mission, i.e., a duration or demand for the performance

response an action or set of actions that are directed to recovering from an initiator or otherwise undesired event

response time the time to a specified response from the time of initial opportunity to respond

risk (1) the chance of a loss or damage, or (2) the frequency of an undesired consequence, or (3) the uncertainty of a hazard

rule-based behavior a (hypothesized) mode of behavior that amounts to following situation-action pairs

schema [after Neisser, 1976] a neurological system for aiding perception or cognition

sequence a chain of events that trace an initiating event to a specific consequence

skill an ingrained ability or capacity toward specific action [which may be innate or be learned]

slip (1) an error in implementing a plan, decision, or intention [the plan is correct; its execution is not], or (2) an unintended action

software (1) computer instructions, or (2) any information stored on paper, film, electromagnetic media, etc.

standby a state of a device in which it is statically awaiting to perform its designed function

stress the physiological or psychological reaction to loads, burden, or other stressful influences on people

system a group of entities—hardware, software, people, or their combination—that interact to produce joint behavior that can be measured against some goal or standard

systems engineering (1) the engineering of systems (i.e., almost any engineering), or (2) engineering or engineering analysis concerned primarily with system performance

task a series of human activities designed to accomplish a specific goal

taxonomy a classification or way to classify

time reliability correlation a relationship of the probability of the (failure of) occurrence of an event to the time over which the event could occur

transient (1) an initiator, or (2) any off-normal condition, or (3) just a change

troubleshooting a form of diagnosis associated with identifying the reason for the failure of a device or system

uncertainty a lack in knowledge or a failure in being able to predict a postulated event

vigilance continual monitoring of displays, people, or equipment [technical babysitting]

workload the burden placed on a person due to the job or task characteristics, particularly as it produces physiological stresses [the concept of **mental workload** also exists (Moray, 1979)]

INDEX OF AUTHORS

The following people are cited in the text and can be found in the bibliography.

A
Ablitt, JF 9
Adams, JR 11
Altman, 5
Anderson, H 43
Askren, WB 5

B
Bain, LJ 29, 50
Barlow, RE 3
Bazovsky, I 3
Beare, AN 41, 49, 138
Bell, CR 149
Berliner, C 3, 4
Birnbaum, ZW 3
Bongard, M 15
Bott, TF 28, 41, 49
Broadbent, DE 149
Brune, RL 40

C
Card, SK 36, 108
Chang, NE 9
Cherns, AB 17
Chexal, B 81, 187
Collins, EP 200
Comer, MK 47, 48
Curry, RE 10

D
Dennet, DC 12-13, 15, 18
Dhillon, BS 2
Donovan, MD 47
Drago, JP 199
Dreyfus, HL 10, 14, 186

Dreyfus, SE   10, 14, 186
Duncan, K   11, 18

E
Embrey, DE   7, 48, 67

F
Fleming, KN   9, 32, 110
Fragola, JR   9, 10, 40, 41, 42, 110-111, 118-119, 200
Frankl, VE   14, 15
Fullwood, RR   9
Furst , C   13, 15

G
Gazzaniga, MS   13-14, 18
Gedenko, BV   3
Gertman, DI   47
Goodstein, LP   2
Greene, AE   9, 52, 118
Guttmann, HE   5, 10, 29, 41, 47, 59, 78, 126, 134, 181, 185

H
Hall, RE   10, 42, 46, 110-111, 118, 159
Hannaman, GW   32, 79, 111
Hartley, HO   113
Henneman, RL   22, 149
Hess, RA   18, 19, 20, 22, 149
Hoffer, E   14
Hofstatder, DR   13, 15
Hollnagel, E   111
Husseiny, AA   9

J
Janis, IL   14, 17, 18, 138, 149, 186
Jenkins, JP   5
Jensen, RS   1
Johannsen, G   15, 17, 22
Johnson, WG   71

K

Khaneman, D 15
Kelley, D 14
Kemeny ix, 6, 187
Kincade, RG 5
Kolb, GJ 9
Kukielka, CA 157

L

Lautman 5
Lees, FP 71
Leplat, J 11, 18
Lerner, EJ 15, 22
Lewis, HW ix
Liwång, B 43
Loyd, DK 3
Lipow, M 3
Luckas, WJ 46

M

Macnamara, JT 13, 15, 16
Mann, L 17, 11-8, 149
May, R 15, 17
Maynard, HB 3
McClymont, AS 193
Meister, D 5
Miller, GA 149
Minarick, JW 157
Moray, N 149, 159
Munger, SJ 3, 41, 48
Mycielska, K 36, 133

N

Neisser, U 14-15, 17, 18
Newton, HP 18, 22
Nielsen, D 71
Norman, DA 16, 87
Nozick, R 15

O
O'Brien, JN   42-43, 46

P
Papazoglou, IA   48
Park, KS   2
Parker, JF   5
Pearson, ES   113
Pedersen, O   111
Perrow, C   1, 6, 77, 149
Phillips, LD   48
Poehlman, BW   193
Pribram, KH   14
Prochan, F   3

R
Rasmussen, J   2, 6, 11, 15, 18, 20, 111
Reason, JT   1, 11, 12, 16, 36, 78, 87, 133
Rigby, LK   5
Rogovin, M   ix, 139
Rose, SPR   14
Rouse, WB   15, 17, 22, 149

S
Sandler, GH   3
Schmall, T   ix, 6, 41
Siegel   5
Shooman, ML   2-3, 27
Smith, KN   16
Spettell, CM   43, 46
Stillwell, WG   101
Swain, AD   ix, 5, 10, 29, 41, 47, 59, 78, 110, 126, 134, 159, 181, 185

T
Topmiller, DA   47
Trivers, R   18, 22
Tversky, A   14

V

Van Cott, HP   5
Vesely, WE   75, 194, 196
Villemeur, A   49
vonHerrmann, JL   131
Voska, KJ   42

W

Waage, JM   152
Waller, RJ   149
Weiner, EL   10
Welford, AT   149
West, VR   5
Whitehead, DW   48-49, 53, 112
Wohl, JG   22, 33, 149, 157
Wolf   5
Woods, DD   20, 139, 187
Woodson, WE   3, 4
Wreathall, J   10, 42, 110-111, 118
Wycoff, H   81, 187

INDEX OF TOPICS

The following is a list of topics that played a critical role in this book.

A
Accident Initiator Progression Analysis (AIPA)   110
accident sequence   75, 194-195, 201
accident sequence analysis   194
accident sequence definitions   191
Accident Sequence Evaluation Program (ASEP)   46
aggregation of data   119
AIR (American Institute for Research) Data Store   3
anchor probabilities   103
anticipated transient without scram   46, 53, 57, 123, 131, 145
anticipation   15
Aristotelean Greek tradition   186
artificial intelligence   186
assistant senior reactor operator (ASRO)   140, 142-143
automatic depressurization system (ADS)   151, 131, 133
automating of process control   186
auxiliary feedwater (AFW)   150, 157

B
basic events   75, 197
basic probability   136
basic time reliability correlations   141
Bayesian update   200
behavior mechanisms   20
belief structures   182
Berliner classification   6
biases, accounting for   200
bins   75, 76
Bioastronautics Data Book   5
Boolean equations   75, 197
Brookhaven National Laboratory   41, 42, 43, 111
Browns Ferry Fire   153
burden   19, 22, 32, 66, 149, 150, 154, 155, 156, 157, 158, 159, 122
BWR Owners' Group   168-170, 179

C

calibration of redundant instruments   136
categories of behavior   15
cause-consequence diagrams   71
characteristic response time   116
Chernobyl   71, 90, 142
Chi-squared distribution   200
classification of human performance   3, 6, 11, 81-82, 86, 89, 91, 186
Code of Federal Regulations   153
cognitive model   8
combination of generic and site-specific data   200
command and control burden   152, 182
commissions (action due to misdiagnosis)   144
common cause failure   197
complementary cumulative distribution function   34, 35, 159, 109, 125
component (failure events)   201
component exposure calculation and determination   199
component level   196
component type   199
complexity   138, 140
conditional events   137
conditional probability   62
confidence   137
conflict   22, 101, 122, 127, 138-140, 151, 162, 178-179, 181
conjugate prior approach   200
consciousness   19
containment integrity   195
containment spray system   152, 165, 167, 168, 173, 177
containment system response   196
containment systems   196
contingency training   187
convolution of distributions   160
core damage   196
core damage cutsets   196
core damage frequency   192-193, 196
core damage sequences   194-195
core damage risk   182, 192
crew structure   186
critical safety parameters   139

Index   231

cumulative distribution function   28, 29, 30, 33, 42
cutset, accident sequence   75, 96, 170, 172, 201

D

data cell concept   5, 40
data encoding   199
data identification and collection   199
data windows   199-200
decision analysis   186
decision making burden   151
decision tree   89
dedicated shutdown system   154
degree of burden   158
demand failure   64
density function   33
dependency   62, 63, 136-137
dependency factor   136-137
dependency model of THERP   136
diagnosis or decision making   139-140
diagnostic burden   150, 151
distribution parameters   199

E

Electricité de France   49, 51
Electric Power Research Institute   49, 8-1, 79, 112
elemental events   197
emergency core cooling system   94-97, 151
emergency feedwater system   183
emergency procedure guidelines   173
emergency response guidelines   103
error bound determination   199
error factor   32, 118, 121-122, 199-200
error reduction programs   185
event analysis   83
event headers   194-195
event interpretation   171
event specific TRCs   119
event tree   75, 92, 194
event tree analysis   194

event tree representation   172
exponential distribution   32, 51
external events analysis   190

F
familiarization   64, 85
failure categorization   199
failure logic   65
failure mode   199
failure modes classification   87, 138
failure probability   197, 200
failure rates   200
failure regime   25
failure severity   199
fault trees   75, 93, 96, 98, 194, 196-197
feedwater system   169-170
feedwater transient   174
feedwater trip   174
fine screening   133-138
fire scenario   154
front-line system trees   198

G
Garoña   xi, 86, 165, 168-170, 172-173, 178-179
Gaussian distribution   29, 112, 126
general actions' TRC   119, 121
general diagnosis   140, 178
generic action types   53-55
generic component failure basic events   198
generic data base   199
generic data sources   198
goal conflict   138, 188
GO models   75
group dynamics   186
group dynamics, effects on TRCs   119

H
hardware models   210
hazard analysis   187

hazard function 32, 33
hazards and operational analysis 71
HCR TRC 111
hesitancy 22, 122, 126-127, 138, 142, 158, 162, 178, 182, 187
high pressure coolant injection system 165-167, 169, 173-176
high pressure injection 182
homunculus 13
hot shutdown 168
HRA procedure 79
human behavior 23, 191
human behavior process diagram 19, 20
human cognitive reliability model 110-111, 118
human decision making 17
human engineering deficiencies 134-135
human engineering guide 5
human error 77, 185
human error probabilities 42, 46, 49
human error taxonomy 11
human failure 77-78, 185, 201
human failure event record sheet—mistake 143
human failure event record sheet—slip 144
human failure event taxonomy 79, 89
human failure models 27, 34
human-induced initiators 83, 108
human interaction 73, 15-1, 191
human malfunctions 186
human performance analysis 3
human performance theory 12, 186
human reliability data 5, 191
human reliability model 59, 191
human reliability technology 185, 188
human reliability theory 23

I

identifiable plant state 194
IEEE Guide on HRA 79
influences on human reliability 22, 100-101
information processing system 16
initiating event 194, 201

initiating events analysis   193
initiator frequency data   198
initiator group   194
In-Plant Reliability Data System project   199
Interim Reliability Evaluation Program   9, 111
IREP TRC   111
isolation condenser   165-167, 169, 173-176, 179

K
knowledge-based behavior   15, 16, 20, 118
knowledge-based TRCs   122

L
lag time   120
latent failures   72, 88, 93, 95, 132, 135-137, 167
latent probabilities   137
Lewis report   ix
Licensee Event Report   42
light water reactors   67, 151, 108
linking fault trees   201
log-linear distribution   51
log-median   142
lognormal CCDF   113, 126, 142
lognormal density   112
lognormal family   112, 140
lognormal functional forms   30, 35
lognormal parameters   31, 32, 51
lognormal random variable   125
lognormal TRC   114, 140
log-probability format   41
log-probability paper   30, 41, 51, 112, 116, 140
long-term cooling   170-171
loss of coolant accident   59, 60, 61, 94-96, 103-104, 108-109, 115, 151, 152, 167, 169, 174
loss of emergency feedwater   184
loss of feedwater   107, 156, 183
loss of station ac power   165
loss of subcooling margin   182
low pressure coolant injection system   1651, 167, 173-175, 177

Index 235

M
main condenser   170
main steam isolation valve   46, 53
maintenance unavailabilities   200
management oversight and risk tree   71
man-machine interface   9, 43
manning model   66, 109
measure of central tendency   140
median   32
median response time   140
mental workload   149, 159
MIL Standard of Human Engineering Design Criteria   5
mind/brain dichotomy   13
miscalibrations   136
mistakes   16, 17, 87-90, 131-133, 138, 140, 142-143
modularization process   197
multiple TRC approach   182
Myrtle Beach conferences   ix, 6, 41, 78

N
National Reliability Evaluation Program   47
NAVSEA Human Reliability Prediction System   5
net component failure probability   136
nominal diagnosis model   108-110
nominal diagnosis TRC   113, 126
non-recoverable failures   170
Nordic Liaison Committee for Atomic Energy   43
Nordic Project   43
normal (Gaussian) distribution   29, 112
normalizing factor   33
NREP TRC   111
Nuclear Computerized Library for Assessing Reactor Reliability   47
Nuclenor   165

O
OAT TRC   109-110, 115
Oconee PRA   47, 74, 189, 196
off-normal recovery activities   178
operator action event tree   131-132

operator action trees   xi, 7, 9, 110
operator reliability calculation and assessment (ORCA)   83-84, 126-127, 142-143, 145
ORNL simulator experiments   28, 49, 50, 51, 107, 113, 116, 118, 123
ORCA logic structure   146
outage time distribution   200

P

P-300 EEG wave   15
pace (of an accident)   119-120
performance shaping factors   5, 43, 46, 48, 63, 67
photo survey   86, 173
physiological burden   152
pilot-operated relief valve   150-151, 169, 181
plant deterministic analysis   191
plant risk model   190
plant-specific data analysis   199
plant-specific database   198, 200
plant-specific initiators   194
plant-specific risk model   202
post-initiator human failures   93-95
PRA final report   201
PRA level 1   189-192, 196, 201
PRA level 2   189, 195-196
PRA level 3   189, 196
PRA Procedures Guide   9, 79, 196
probabilistic risk assessment   74-75, 77-78, 81-83, 85, 88, 92, 94, 185, 189-190, 192, 195, 198, 201
probability density   29
probability of slips   135
procedure-based classification   89
procedure talkthrough   169
proximity failure cause   199

Q

qualitative assessment   64-65, 85, 131
qualitative screening   81
quantification system   142
quantitative assessment   64-65

R

Rasmussen model 7, 11
Rasmussen taxonomy 6, 11
recommendations for risk reduction 202
reactor coolant system 157, 182
reactor core isolation cooling system 53, 57
reactor pressure vessel 167-168
Reactor Safety Study ix, 5
reactor shutdown coolant system 165, 167
recirculation failure event 104
recoverable failures 170
recovery 84, 89, 93, 97-98, 160, 167, 169,. 172, 201
recovery curves 140
recovery events 135
recovery failures 201
recovery TRC with hesitancy 133
recovery TRC without hesitancy 126
redundancy 61, 63
reliability 3, 23
reliability analysis 25, 59, 61
reliability block diagram 75
reliability engineering 3
residual heat removal system 51, 152
response curves 140
response time 27
response time correlations 32
response time distribution 140
response time parameters 31
response time probability regimes 42
risk analysis 71-74
risk assessment models 189
risk-benefit estimate 202
risk criterion 192
risk management 201
risk-significant sequences 86
RISØ National Laboratory 11
RMBK-1000 86
root cause evaluation 198
rule-based behavior 7, 20, 118

238   Human Reliability Analysis

rule-based events   131, 133, 178
rule-based mistakes   140
rule-based with hesitancy   133
rule-based TRCs   122
rules   20

S

Sandia human error data bank   5
Sandia Human Reliability Handbook   185
Sandia National Laboratories   51, 112
Santa Maria de Garoña   165
schemata   18
scoping model   201
screening strategy   133-134, 137
screening value   133, 135
senior reactor operator   142
sequence class   170
sequence cutsets   167
sequence phase classification   88
shift technical advisor   184
short-term memory   149
skill-based behavior   20, 118
SLI calculation   184
SLI calculus   103
sliding time windows   156
slips   16, 17, 36, 87-90, 131-134, 143
"smart" fires   154
solution rate   32, 122-123
sources of burden   150, 153
spatial reversals   134
standby liquid control system   57, 151, 179
station blackout   53, 56, 93, 122
steam generator tube leak   116
stereotype capture   134
stress   19
structured interview technique   200
success criteria   191, 195-196
success likelihood index   101-105, 123, 127, 142, 179, 183, 185
success likelihood index methodology   67, 99, 101, 103-105, 126, 131, 142

support system fault trees   198
support system models   194
surrogate data   185, 198, 200
symptom-oriented emergency procedures   139
Systematic Human Action Reliability Procedure   10, 79
system failure events   194, 196
system fault trees   167, 194
system interactions   194
system reliability analysis   73, 75
systems modeling and analysis   196, 198

T

task analysis   60, 169
task capture   27, 37
task units   5
technical support center   184, 187
THERP   ix, 5, 7-10, 47-48, 59, 61-62, 65-67, 78, 82-83, 87-88, 90, 99, 107, 109-110, 113, 131-132, 134, 140-141, 159, 160, 182
THERP TRC   127, 131
threat stress   108
Three Mile Island   ix, 6, 10, 67, 107, 110, 141, 150-151, 181, 189-190
threshold probability of failure   161
time-dependent stochastic process   140
time reliability correlation   9, 10, 34, 36, 43, 48, 53, 58, 82, 90, 108-115, 118, 123-126, 129, 131-132, 140, 144, 158-161, 184
time reversals   134
top-down approach   192, 195
top logic   194-196
top of active fuel   177
torus cooling   167, 170-171
transient sequences   166
TRC parameters   57
TRC system   107, 124, 129, 141-142

U

unavailability data   199
unavailability parameters   200
unavailabilities   197

uncertainty   71, 115, 125-126, 131, 137-138, 149
unrecoverable slips   178

V
validation of HRA   82
vigilant behavior   37, 52
volatility of data   40

W
walkthroughs   86, 139
WASH 1400   5, 82, 108, 167-168, 189-190
Weibull distribution   32, 51, 111-112, 123